About Island Press

Since 1984, the nonprofit organization Island Press has been stimulating, shaping, and communicating ideas that are essential for solving environmental problems worldwide. With more than 1,000 titles in print and some 30 new releases each year, we are the nation's leading publisher on environmental issues. We identify innovative thinkers and emerging trends in the environmental field. We work with world-renowned experts and authors to develop cross-disciplinary solutions to environmental challenges.

Island Press designs and executes educational campaigns, in conjunction with our authors, to communicate their critical messages in print, in person, and online using the latest technologies, innovative programs, and the media. Our goal is to reach targeted audiences—scientists, policy makers, environmental advocates, urban planners, the media, and concerned citizens—with information that can be used to create the framework for long-term ecological health and human well-being.

Island Press gratefully acknowledges major support from The Bobolink Foundation, Caldera Foundation, The Curtis and Edith Munson Foundation, The Forrest C. and Frances H. Lattner Foundation, The JPB Foundation, The Kresge Foundation, The Summit Charitable Foundation, Inc., and many other generous organizations and individuals.

The opinions expressed in this book are those of the author(s) and do not necessarily reflect the views of our supporters.

Advance Praise for
The Power of Existing Buildings

"Today, most of our old building stock requires updating to reflect the current needs and demands of today's clients, as well as our new climate reality. This book offers a sound approach to tackling this task, based on the building science practices endorsed by the North American Passive House Network (NAPHN). It includes a viable financial pathway to support what must be seen as an evolution; the updating of old systems to allow the charm and character inherent in old buildings to remain viable for future generations."

—Bronwyn Barry, RA, CPHD, NAPHN *Board President*

"Buildings make up 80 percent of the City of Pittsburgh's carbon footprint. If we are going to make meaningful progress in reducing carbon emissions, we need to address the backlog of opportunity within our existing building stock. *The Power of Existing Buildings* provides a roadmap for building operators, policymakers, real estate developers, and anyone who is looking to create higher performing buildings. These techniques and strategies are key components of our game plan to push buildings to net zero."

—Grant Ervin, *Chief Resilience Officer*, City of Pittsburgh

"Owners lack confidence that their investments in existing buildings will lead to improved performance in operations, so quite often they do nothing. *The Power of Existing Buildings* lays out the building science and technology-based tools available today that are essential to fully integrating design, construction, and operations, ultimately increasing an owner's return on investment. The more integration on a project, the higher the return."

—Bob Berkebile, *Principal Emeritus, bnim*

"Internationally, industry leaders are successfully and affordably delivering deep energy retrofits of existing buildings. Many countries either have, or soon will have, building codes requiring such retrofits; yet how they are achieved remains unknown to many. *The Power of Existing Buildings* is a compelling synthesis of the fundamental principles enabling the successful delivery of such projects, and is essential reading for owners, project teams, and policy makers."

—Rob Bernhardt, *Chief Executive Officer*, PassiveHouse Canada

"A brilliant intersection of design, technology, building science, and operations, this definitive work is the roadmap to zero in existing buildings. This is a must read."

—Laura Nettleton, *Founder and Architectural Coordinator*, Thoughtful Balance

THE POWER OF EXISTING BUILDINGS

The Power of Existing Buildings

Save Money, Improve Health, and Reduce Environmental Impacts

Robert Sroufe, Craig Stevenson,
Beth Eckenrode

Washington | Covelo | London

ISLAND PRESS is a trademark of the Center for Resource Economics.

Library of Congress Control Number: 2019938573

All Island Press books are printed on environmentally responsible materials.

Manufactured in the United States of America
10 9 8 7 6 5 4 3 2 1

Keywords: Architecture 2030 Challenge, AUROS360, building envelope, building reuse, building simulation, carbon neutrality, efficiency, energy modeling, energy-use intensity (EUI), EnerPHit, evidence-based performance, existing buildings, future-proof buildings, green building, high-performance building, historic buildings, indoor air quality (IAQ), indoor environmental quality (IEQ), integrated management, integrated sustainability dashboard, Internet of Things (IoT), natural order of sustainability, open-integrated network, Passive House; Pittsburgh, Pennsylvania; single pane of glass, smart building infrastructure, social cost of carbon (SCC), sustainable buildings, sustainability in the built environment, sustainability modeling, UN Sustainable Development Goals, whole-building performance, Zero-Energy, zero net energy building.

Contents

Foreword   xi

Acknowledgments   xv

Introduction: Why Should I Think about Retrofitting
                My Building?   1

Chapter 1:   My Building Has High-Performance Potential   9

Chapter 2:   Where Do I Start?   35

Chapter 3:   The Importance of a Project Plan: Every
             Building Needs One   51

Chapter 4:   Can I Afford This?   65

Chapter 5:   The Building Envelope Holds the Key   85

Chapter 6:   How Realistic Is Zero-Energy for an
             Old Building?   107

Chapter 7:   Operating Buildings for Maximum Benefit   123

Chapter 8:   Case Studies   153

Chapter 9:   Existing Buildings Can Save the World   169

Appendix 1: Building Your Plan: Project Development
            Homework   189

Appendix 2: Critical Resources on Existing Buildings   194

Notes   199

Foreword

The availability of large quantities of fossil fuels has fundamentally shaped and changed the way people live during the past 100 years. In industrialized countries, an average person "consumes" more than fifty times (fifty!!) the energy from commercial supply than his or her ingested energy for metabolism. The vast majority of today's energy supply (2019) is still based on fossil fuels. In other words, it is obtained by burning the carbon extracted from coal, oil, and gas storage sites that have taken hundreds of millions of years to form. The organisms that formed these storages originally took this carbon from the atmosphere where it was contained in the form of carbon dioxide. Through our modern way of life, we are now re-introducing this carbon back into the atmosphere within the very short period of time of only a few decades. We are effectively turning the planet back into a greenhouse and thereby creating conditions under which an advanced civilization—with soon 10 billion people—can no longer exist in dignity. A very distressing outlook.

The fact is, most people feel very comfortable with the lifestyles they have gotten used to. They don't want to compromise on large living spaces, work-saving machines, easy communication, and quick, long-distance mobility. So, is there a way out of the current distraught situation?

Yes, there is—by switching to a sustainable, circular economy based on renewable resources. In the energy sector, this transition has two main components: Firstly, the stringent improvement of energy efficiency and secondly, the use of renewable energy sources. The fact that there is immense potential for improving the level of sufficiency and efficiency is evident simply from looking at the vast and wasteful amounts of fossil fuels that are currently being consumed. Diminishing unnecessary losses, and thus increasing overall efficiency, leads to a significant reduction in total energy demand. Only if the energy needs are substantially reduced, the limited globally available renewable resources (such as photovoltaic and wind power) will suffice to provide the desired standard of living.

About forty percent of today's final energy consumption (i.e., everything purchased by end consumers) is used to provide services in buildings—which, in turn, is clearly dominated by the needs for heating, with well over seventy percent in Europe and North America. The fundamental reason for these high-energy needs are losses through the envelope (i.e., the heat that escapes the building through walls, roofs, floors and windows). These heat losses can be drastically and effectively reduced simply by using components and applying detailing practices that have been developed and gained more and more market recognition over the past thirty years, such as thermal insulation, triple-glazed windows, thermal bridge reduction, and mechanical ventilation with heat recovery. This has been successfully demonstrated by many built projects around the world—newbuilds and retrofit projects.

The real challenge in terms of increasing efficiency is the existing building stock, which was mainly built during times when energy seemed to be abundant and cheap, and when efficient components were not available on the market. The good news being that existing buildings can be successfully retrofitted with the above-mentioned state-of-the-art Passive House components. If well planned, their heating energy consumption is reduced by a factor of three to five. This, in turn, makes it possible to completely electrify the services—especially the heating—and thereby enable a transition to a renewable, friendly energy supply that no longer needs to rely on fossil energy resources.

It is important to acknowledge the fact that individual building components are often retrofitted at different times, according to their own expired lifetime. These events are the exact opportunities to not just replace old components, but to improve their efficiency at the same time. If the improvements are not done at this time, it will be a lost opportunity. Therefore, it is really important to make the future-proof decisions at each of these refurbishment steps: "If you do it, do it right."

How this can be planned, implemented, and used successfully is covered in this book. Well-planned, deep energy retrofits ensure cost-effectiveness, longevity, and comfort.

All those who envisage a sustainable future for our common cultural heritage should read this book carefully. Only by spreading the knowledge and the uptake of sustainable solutions as quickly as possible can we make the changes and progress that is absolutely necessary if we want to preserve the blessing of modern prosperity for future generations.

Prof. Dr. Wolfgang Feist
Founder of the Passive House Institute
and laureate of multiple environmental awards

Acknowledgments

The information in this book is the product of years of work, shared learning with my coauthors (without them this book would not be possible), the showcasing of best practices, and the development of knowledge across teams to gain insight into existing building opportunities.

Duquesne University and my graduate students have all contributed to my thinking, and this book, through our annual building design competitions and their work within an integrated, top-ranked MBA curriculum. These change agents inspire innovation as they analyze business enterprises and the built environment; understand drivers and risks; identify problems and opportunities; evaluate return on integrated investments from alternative courses of action; and value short- and long-term environmental and social prosperity with goals of integrated bottom-line performance.

I need to thank Research Fellows Kevin Dole and Laura Jernegan for their help with draft versions of this book. There are a multitude of colleagues from across industry and academia who helped spur my thinking and answer questions about the changing performance frontier of the built environment and whole building performance. Many of them who have directly impacted my thinking are from the Green Building Alliance; Conservation Consultants, Inc.; and the International Living Future Institute, its Living Product Hub; the Rocky Mountain Institute; and Carnegie Mellon University's School of Architecture. I cannot mention all individuals here, but many are mentioned through this book. Due to these prior efforts, readers can independently learn about, apply, and reflect on proven models and methodologies while honing their skills in project teams and set goals for improving their buildings.

I need to thank my family for supporting and enabling the development of this and other books on our own path toward sustainability. My wife for putting up with my work to improve our home to make it a net zero energy building. To my daughters, I look forward to seeing you grow into the building, environmental, and social advocates for change I know you will be.

To the readers of this book, thank you for reading our work. I hope you get to realize the power of your own buildings as we create the changes necessary to thrive in a sustainable society.

Robert Sroufe

I would like to thank my coauthors, Dr. Robert Sroufe and Beth Eckenrode, for their generous collaboration of insight, energy, and time on this book. The concepts of this book began during years of collaboration with Robert at Duquesne University's Globally Ranked Sustainability MBA program. Beth and I are partners at the AUROS Group; if you ever want to test a partnership, just write a book. C.S. Lewis said, "Integrity is doing the right thing even when no one is watching." The tireless effort to get it right in this book while maintaining schedule and structure reinforces what I have always known about Beth's personal drive. We are both grateful to Bhakti Dave, AUROS Group, for her keen insights and support with the book.

I am forever indebted to my wife, Liesl, for her patient understanding, along with my sons—Tyler, Alexander, and Trevor—who are all old enough now to know that it's time to act and reverse climate change. My parents, James and Sandra, instilled in me my entrepreneurial and innovative spirit at a young age. I sincerely hope that I was able to pass on these qualities to my sons.

I am grateful to Brenda Morawa and her IES VE team for their availability to pressure-test our many early ideas. Similarly, I thank Mark Benninger for his patience and understanding as we continuously challenge his OSIsoft team with approaching the problem of data acquisition in the built environment from different angles.

Laura Nettleton has proven to be a great friend who has consistently and, without fail, always led from the front by proving that we can build better buildings without spending a premium. Laura was a great resource during the development of this book. I would also like to acknowledge Rick Avon and Nathan St. Germain as high-performance-building leaders who are always available to help solve difficult envelope problems. On an international level, I am proud to work with Bronwyn Barry, Ken Levenson, and Rob Bernhardt at the North American Passive House Network to promote the Passive House standard. You never forget those people who helped you in the early days; and for me that is Stefani Danes, Carnegie Mellon University. Stefani is simply one of the best people I know, and she will always have a special place in my heart.

Craig Stevenson

I thank Craig Stevenson and Dr. Robert Sroufe for convincing me to write this book. The mash-up of three very different backgrounds and experiences worked only because the competencies and leadership qualities of these two men are unmatched. I am proud of the work Craig and Robert have done to find actionable and meaningful solutions to society's most pressing problems.

The inspiration for all that I do comes from my family. My husband, Tim, reminds me that we win when others win. He is caring and unselfish, and I am a better person for his love and support. Together, we have two exceptional sons, Baxter and Harrison. As they begin their adult journeys in life, I wish for them a safe and responsible world led by unselfish people who know that elegant solutions to complex problems require innovation and collaboration between governments, foundations, nonprofits, and for-profit companies. My parents, Richard and Mary Hottinger, have their fingerprints all over my life and my accomplishments. They taught me the enduring lessons upon which I parent, lead, and live.

Thanking everyone else is an almost impossible task. I thank Dimitri Jeon at the Dow Chemical Company for modeling the leadership to which I aspire. I thank Jon, Peter, and David Huntsman at Huntsman Corporation for insisting that we "do good while doing well." It was years later that I fully understood how much the Huntsman family values touched me and informed my development. I thank Jeff Lipton and Chris Pappas at NOVA Chemicals for gleefully throwing me in the deep end always believing in my capabilities. I thank Bill Johnson for his tutelage at the H.J. Heinz Company. I had the privilege to learn from a host of accomplished business leaders, including my father, and I appreciate the investments they each made in me.

To my friends, personal and professional: Kelly, Tony, Cheryl, Mark, Mary, Andy, Jamie, Bhakti, Laura, Gunjan, Celia, Lori, Sharon, Antonio, Kristen, Bill, Dina, Jack—I borrowed from each of you important qualities and thank you for the impact you've had on me.

Beth Eckenrode

Introduction:

Why Should I Think about Retrofitting My Building?

The greenest building is the one that is already built.
— Carl Elefante

C ARL ELEFANTE WROTE THESE WORDS over a decade ago when he was the director of sustainable design at Quinn Evans Architects in Washington, DC. He knew then that no amount of new green construction will get us where we need to go if we ignore existing buildings, adding that "four out of every five existing buildings will be renovated over the next generation while two new buildings are added."[1]

There is a total area of over one trillion square feet in the world's existing buildings today. Existing buildings are among the largest contributors to greenhouse-gas emissions (GHGs). If not designed, retrofitted, and managed properly, they can be very wasteful, costly to operate, and unhealthy.

Recognizing the potential of your existing building now comes at a critical time in history for the built environment in light of global goals for more-sustainable structures and addressing climate

1

change. It's a given that most of the buildings we have right now will still be in use in 2030. That year in the future is a milestone date for carbon neutrality, a goal emphasized most recently by the Intergovernmental Panel on Climate Change (IPCC) and the basis for the work of Architecture 2030.[2]

In the United States, there are over five and a half million existing buildings over fifty years old, all of which have various areas in which performance needs to be improved. Old buildings—and some new ones—often contribute to health problems such as asthma and allergies, due to poor indoor environmental quality.

Data from the US Energy Information Administration shows that buildings are responsible for almost half (48 percent) of all greenhouse-gas (GHG) emissions annually. Sixty-six percent of all electricity generated by US power plants goes to supply the building sector. Both the generation and procurement of renewable energy and a significant increase in the rate of existing-building energy-efficiency renovations are required to meet the emissions reduction targets set by the Paris Agreement.[3] Yet, currently, building renovations affect only 0.5–1 percent of the building stock annually.

There are a number of reasons to renovate an old building:

- *Cost savings*: Old buildings generally leak air, resulting in high energy costs. New mechanical systems, sensors, and meter technology can be added to a building when it is renovated to help dynamically improve its performance.
- *Improved indoor environmental quality*: Occupants today want indoor environmental quality that includes better air quality, thermal comfort, and lighting, as well as fewer noise distractions. In addition to improving employee health, better indoor environmental quality improves worker productivity.
- *Lower carbon footprint*: Reusing an existing building can help building owners and government organizations reach goals of reducing the use of fossil-fuel-based energy.

Each year, the US Green Building Council (USGBC) reports on the trends in green building from the previous year.[4] For several years, renovations have been at the top of the list. In 2016, USGBC noted that one part of its Leadership in Energy and Environmental Design (LEED) rating system, for Building Operations and Maintenance (also called LEED-EB or EBOM), was the most-used rating system in the top ten states for LEED-certified buildings. That year, LEED-EB represented about 53 percent of the total certified square footage. Benefits include lowering operating costs, providing documentation of quality assurance, educating occupants about sustainability, and creating higher-value points of sale. Recent rollouts of the latest LEED upgrade, Version 4.1, started with a focus on existing buildings, and the renovation trend is expected to continue. The World Green Building Trends SmartMarket Report predicts that the number of industry respondents who expect to do the majority (more than 60 percent) of their projects green will jump from 27 percent in 2018 to almost half (47 percent) by 2021.

We can see that some professionals are getting better at finding the potential value in existing buildings. Yet current building-design and -construction practices make it challenging to determine a building's potential in project planning and goal setting. How many times have you heard a design team tell an owner such as a school district that it is going to be so expensive to renovate their school that they might as well build new? "Wow! That's awesome! I get a new building for the cost of improving an old one?" Who wouldn't want that!

Here's the problem with that kind of thinking: it does not consider embodied energy, does not look at total costs or whole-building solutions, and needlessly ends the lifespan of the existing building.

Let's instead flip this "first-cost" argument. For the same cost as building a new building, I can restore the solidly built building I already have, reduce energy consumption, and improve indoor environmental quality all without incurring the costs and energy consumption of demolishing the old building and replacing it with a new building. In a study called "Embodied Energy and Historic Preservation: A Needed Reassessment," Mike Jackson finds that if a building is demolished, partially salvaged, and then replaced with a new energy-efficient building, it would take 65 years to recover the energy lost in demolishing the building and reconstructing a new structure.[5] Sixty-five years? That's longer than most buildings survive.

Unfortunately, the more exciting and appealing recommendation is always going to be a new building; the assumption most teams make is that it will be easier and cheaper to build a high-performing new building than to restore an old building to high-performance standards.

The purpose of this book is to convince you that this assumption is wrong, and we will empower you with the proof to challenge that thinking. Modern passive-building science as well as current smart-building and simulation technologies enable project teams to make old buildings look and perform just like new buildings.

What is growing quickly in the most enlightened circles of sustainability and building performance is the notion that the "greenest" buildings in the world, by default, will be existing buildings restored to ultra-low energy consumption and ultra-high indoor environmental quality. Today, building lifecycle analyses show that the greenest restored building always make more sense than the greenest new building.

This change in learning curves and mindsets regarding existing buildings has not come overnight (see fig. 0-1). Building on a conservation focus since the 1970s, decision makers have, without performance feedback loops, progressed through theoretical and prescriptive standards on to performance standards and first costs. By first costs, we mean the initial cost of the project without looking at lifecycle costs and impacts on whole-building systems. Since 2007, we have crossed the proverbial existing-building chasm with cur-

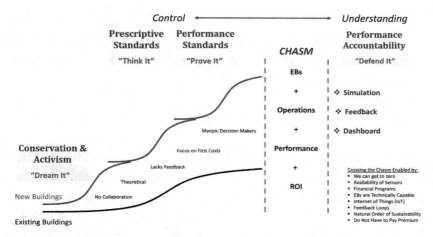

Figure 0-1. Existing Building Learning Curve.

rent learning curves, combining existing-building operations, performance, and returns on investment, to reach a new state of performance accountability. We now have added benefits from simulations, technology-based feedback from buildings, and real-time dashboards enabled by smart building infrastructure, sensors, meters, Internet of Things (IoT), and numerous frameworks for decision making that show we can afford to invest in existing buildings and, at the same time, the people who work and live inside them.

Why Should I Keep Reading This Book?

This book is the result of a collaboration between industry and academia to provide a logical pathway for owners of existing buildings to optimize every dollar they invest in those buildings. We agree that, given the massive real estate that is already built, establishing and reaching each building's theoretical optimum performance level is by far the most sustainable and cost-effective approach to lessening the built environment's impact on the planet and human health.

A key takeaway from this book and our approach to existing buildings is that technologies and building-science strategies that already exist today can transform old buildings into new buildings without spending a king's ransom.

We start with chapter 1 with an overview of the built environment, typical learning curves, current applied systems thinking, and emerging technologies that enable high-performance buildings. Subsequent chapters provide guidance on what to look for—collaboration and planning, representing owners' and occupants' interests, performance advocacy, sustainability certifications, the business case for existing buildings, operational whole-building modeling, and smart building infrastructure when thinking about restoring existing buildings. We review deferred maintenance, equipment lifecycle, renovation triggers for projects, and how to develop a whole-building plan. We explain how to manage a process of transformation and provide case studies of successful renovations. We also look at how to set goals for Zero-Energy and superior indoor environmental quality, and show you how to ensure that your building lives up to, or even exceeds, its potential.

Each chapter lays out a stepwise approach to realizing the full potential of your building—baseline data and the Internet of Things (IoT), planning practices, collaboration, the business case for renovating buildings, how the envelope should be a key aspect of your project, setting goals for zero-energy buildings, and how to operate for high performance. Existing buildings have extraordinary untapped potential to increase revenue and improve overall equity value if the right decisions are made at the right times. This book will explain how to make old buildings perform like new ones.

It is possible to do better with your existing buildings, and the information in subsequent chapters will help you to realize a plan.

This book will help you know where to start, how to think through financial options, and how to realize your goals of net-zero and sustainable development.

This book is intended to refute outdated assumptions about old buildings, expose the full potential value of buildings, and empower building owners to make more holistic decisions based on lifecycle costs. High-performance buildings do pay. It's possible to get out in front of global issues, coming policy changes (such as Passive House, 2030 Challenge, and Zero-Energy), and develop the plan for any existing building to perform like a new building. For those of you in corporations, your existing buildings can and should be part of your corporate sustainability reporting. Even incrementally improving the performance of your existing buildings will differentiate your company from competitors in your industry, but this advantage will not last long. As every hockey player has heard, Wayne Gretzky explains that to win the game you must skate to where the puck *will be*, not where it is now. If you look too incrementally to improve the performance of your existing buildings, you miss out on the full value and potential of existing buildings.

The remainder of this book is dedicated to empowering enlightened owners to build and/or renovate the highest-performing building at the lowest possible costs. We advocate for using building science to ensure that decision making is done in a proper and logical sequence. Such an approach is the secret to providing building owners with the tools and confidence to radically change building performance. For our purposes, when we refer to building performance we are always referring to the reduction of energy consumption *and* the dramatic improvement of indoor environmental quality. This is not a tradeoff approach to improving buildings; instead, we take a synergistic approach that considers investment, data, technology, and aligned teams and decision makers. Using our approach, we know that building owners never have to sacrifice energy conservation to get indoor environmental quality or indoor environmental quality to get energy conservation.

Chapter 1:

My Building Has High-Performance Potential

I spent a great deal of money on this building, [but] I don't think it's performing. I can't see improvements in my utility bills and I don't feel like my building is any healthier. Something needs to change.

— Typical Existing-Building Owner

IF "SOMETHING NEEDS TO CHANGE" resonates with you, then you probably need to read this book now. You will see that it is possible today to make old buildings perform like new ones without paying a premium in construction costs. The primary enablers are modern building science, sophisticated data analytics, and the harnessing of a vast offering of Internet-of-Things (IoT) capabilities using a simple and elegant methodology. Our thesis that an existing building is the greenest building stands in contrast with the notion that the greenest building is the brand-new, highest-scoring, LEED-Platinum award winner with visible solar panels on the roof, or a living building, or one that has a perfect bioswale next to multiple forms of transit. We are redefining "green" and what it means to existing buildings around the world.

We go into this book knowing how difficult it is to be a sustainability advocate in today's environment. Knowing what is possible and how much better buildings can perform makes it challenging to put up with underperforming buildings or resistant project partners—you cannot unknow what you know. The time you spend with this book will help you convince other people that improving the performance of existing buildings is worth it—financially, socially, and ecologically.

Throughout the book we touch upon four key themes: (1) the complexity of transforming existing buildings into high-performance buildings, and the need for a holistic solutions; (2) how technology is driving the move toward the delivery of performance during building operations; (3) the difficulties of prioritizing investments that use advanced passive-building science; and (4) the value of learning from what has been accomplished, drawing examples from our research and experience.

Owners of existing buildings today have a growing, giant hairball of an issue—"a tangled, impenetrable mass of rules, traditions, and systems all based on what worked in the past—that exercises [an] inexorable pull into mediocrity."[1] Building owners can see and feel this "impenetrable mass" and the resulting "pull into mediocrity," but most have no idea how to change. There is a general feeling of paralysis among existing-building owners such as city managers, school superintendents, developers, university presidents, and hospital presidents. Decades of putting off retrofits or simply replacing systems in kind are the primary causes of the paralysis. Leaders are also overwhelmed by rapidly changing technology and building standards. Most organizations don't have the funds to throw at a problem in the hope that results will improve. They need guidance.

Project team members (e.g., clients, building owners, developers, consultants, architects, contractors and subcontractors) intend to give the owner what he or she wants, yet often there are a number of challenges to achieving real building performance results in operations. It's not about intentions; it's more about the process of goal alignment. There are many challenges to reaching alignment of goals on a project. First, the owner's goals are not always clearly identified.

Second, the construction process can be a series of disconnected handoffs from architects to engineers to construction managers to builders, resulting in project teams that are not integrated or aligned on specific goals. Third, there is typically a critical lack of empirical evidence, outside of construction costs and fees, to guide owners' decisions between building performance and costs during the construction project life cycle. Finally, there are very few, if any, building performance measurement and verification methods able to prove to owners on day one of operations that they got what they paid for in terms of building performance.

Our intent is not to look backwards and place blame, but to find the fundamental reasons for these problems and then provide achievable and affordable "how-to" strategies to extract full value from the renovation of existing buildings. To successfully convince owners to invest in renovating existing buildings, they have to believe the destination is worth the journey. Can owners derive enough value from existing buildings to justify the risks of investment, both financial and reputational? Finding the best way to convince building owners that investing in change is possible requires a collaborative approach.

Next, we look at managing change and innovation through collaboration.

Change and Innovation Require Collaboration

It's widely understood that people, instinctively, don't like change—nobody likes to see their cheese moved. Most links in the construction value chain are highly commoditized, extremely competitive, with little to no product differentiation. While there are sustainability certification programs available (see box 1-1), given the number and complexity of them, how is any owner, developer, or project team expected to navigate the world of sustainability programs in an industry where a week-long delay in the design or bid process could result in the loss of a project?

Box 1-1. Select Building Certification Systems

International Living Building Challenge (living-future.org)
The goal of LBC is to encourage the creation of a regenerative building environment. The challenge is an attempt to raise the bar for building standards from merely doing less harm to actually making a positive contribution to the environment. LBC helps owners create spaces that reconnect occupants with nature. Specifically, they recognize those who create buildings that generate more energy than they use, capture and treat all water on site, and use healthy materials.

Leadership in Energy and Environmental Design (LEED) US Green Building Council (new.usgbc.org/leed)
LEED encompasses ten ratings systems for the design, construction, and operation of buildings, homes, and neighborhoods. To become certified, contractors must document certain details for the construction and commissioning of a building. In LEED version 4, certification requires a project to aspire to reduce energy use by at least 5 percent of ASHRAE 90.1.

Passive House (passivehouse.com and phius.org)
Passive House is a rigorous standard for energy efficiency in buildings, seeking to reduce ecological footprints. Passive House results in ultra-low-energy buildings that require little energy for space heating or cooling.

Passive House (Phi) EnerPHit
EnerPHit is a Phi Passive House program for certified energy retrofits for existing buildings. Reductions in heating energy demand can be up to 90 percent by using improved thermal insulation, reduced thermal bridges, improved airtightness, high-quality windows, ventilation with heat recovery, efficient heating and cooling generation, and use of renewable energy sources.

We advocate for an approach we call the Natural Order of Sustainability (see box 1-2), which is an energy consumption and indoor environmental quality methodology that treats buildings as a living organism (also known as a biophilic approach). It promotes a *passive first*, *active second*, and *renewable last* strategy, which ensures that the most enduring systems of a building are optimized for performance first. By maximizing the benefits of passive systems first, the size and cost of subsequent systems like heating, ventilation, and cooling (HVAC) and/or renewables is reduced.

Box 1-1. *continued*

RESET Air (reset.build)
The RESET Air certification is a performance-based building standard that specifies air-quality standards, air-monitor equipment and deployment, and air-quality data management. RESET Air is broadly accepted as the most aspirational air-quality standard and serves as a reference for most of the other green building certification programs and international organizations.

WELL Building (wellcertified.com)
The WELL Building Standard is a performance-based system for measuring, certifying, and monitoring features of the built environment that impact human health and well-being through air, water, nourishment, light, fitness, comfort, and mind. The WELL Building standard explores how design, operations, and behaviors within the places we live, work, learn, and play can be optimized to advance human health and well-being.

Net-Zero-Energy Buildings
A net-zero-energy building (NZEB) is one that produces as much energy as it uses over the course of a year. The metrics combine exemplary building design to minimize energy requirements with renewable-energy systems that meet these reduced energy needs.

The Department of Energy (DOE) and the National Renewable Energy Laboratory (NREL) have led most of the work on net-zero-energy buildings to date. Regardless of the metric used for a zero-energy building, minimizing energy use through efficient design should be a fundamental criterion and the highest priority of all NZEB projects.

Being a change agent isn't easy, but it is precisely the way to achieve differentiation and set yourself apart from your competition. You do have to be ready for the skeptics and folks who say, "owners are weary of all the hype around sustainability" or "we can't afford green strategies" or our favorite, "investing in the envelope of an existing building never pays." The key to getting past skeptics and traditionalists is to talk less about designing a high-performance building and focus more on operating one. The problem all sustainability advocates share is that we never identified the end game. If you line up 100 sustainability advocates and ask them to define high-performance buildings, you will most likely get 100 different answers.

Box 1-2. The Natural Order of Sustainability

**Sustainability Planning Methodology of Passive First /
Active Second / Renewables Last**

Passive First
Maximizing passive strategies (i.e., insulation, envelope, air barriers, thermal bridges, shading, windows and doors) first will reduce loads for heating and cooling systems, thereby requiring smaller and more-efficient active solutions for mechanical systems.

The Passive House standard is the most rigorous set of design principles based on building science used to attain a quantifiable and ambitious level of energy efficiency within a specific quantifiable comfort level. Passive House sets the performance standard at approximately 14 kBtu/sf/yr on the basis that every functioning building requires some level of energy to operate. Passive House's philosophy is simple: "maximize your gains and minimize your losses" through climate-specific building science. Passive House has identified the mathematical limits of diminishing returns for envelope performance (see passivehouse.com and phius.org). A passive building is designed and built in accordance with five building-science recommendations:

1. Climate-specific insulation levels with continuous insulation throughout its entire envelope
2. Thermal-bridge-free connections for all building-envelope sections
3. High-performance windows (double or triple-paned windows, depending on climate and building type) and doors
4. Airtight building envelope to prevent infiltration of outside air and exfiltration of indoor conditioned air
5. High-efficiency heat and moisture-recovery ventilation

A comprehensive systems approach to modeling, design, and construction produces extremely resilient buildings. Passive-design strategy uses highly durable material solutions like fenestration, insulation, air barrier membranes, and cladding that have a long use life even in extreme weather conditions. As a result, passive buildings offer tremendous long-term benefits in the form of energy efficiency and indoor air quality. Passive building principles have been successfully applied to all building typologies, from single-family homes to multifamily apartment buildings, offices, hospitals, schools, and skyscrapers.

Some cities, such as Brussels and Dublin, have introduced Passive House criteria—not certification—into their building codes and have achieved transformative results in the energy performance of new construction. As a result, Brussels now demonstrates a large downward trend in GHG emissions, making it a world leader in energy conservation in its building stock.

Box 1-2. *continued*

Active Second

Implementing passive load-reduction strategies will reduce the size of the active systems and mechanicals required to ventilate, heat, and cool buildings. Design loads in a passive-house building are drastically lower because of the focus on the envelope and insulation, extreme airtightness, and superefficient windows. In simple terms, the building will be easier and cheaper to heat and cool, and the air quality will be better.

Strategies to reduce energy consumption for heating and cooling are most effective when mechanical equipment is decoupled. Logically, planners will optimize passive space-conditioning solutions as a core mixed-mode design strategy. Common passive space-conditioning solutions include an independent balanced mechanical ventilation system with heat and moisture recovery and preconditioning. This strategy will maximize a constant and filtered fresh air supply. Remaining peak loads can then be further mitigated by implementing highly efficient active heating and cooling systems.

Building-enclosure air-tightening means that moist, dirty air isn't leaking into the building's interior space from exterior sources. A constant flow of fresh filtered air flushes the living space without pulling in hot, cold, or wet air that the HVAC system must then condition.

Planners are challenged to manage internal loads and plug loads with efficient appliances, HVAC, plumbing and lighting systems that minimize sensible and latent loads and internal gains. If everything is done properly to this point, a new building will be designed to perform at approximately 14 kBtu/sf/yr, and an existing building will be designed to perform at approximately 20 kBtu/sf/yr, making them both perfectly positioned to reach 0 kBtu/sf/yr—Zero-Energy.

Box 1-2. continued on page 16

Instead of following the traditional "loud voices" in the room, building owners, with the support of building performance advocates, have an opportunity to (1) establish their performance goals based on their own building(s); (2) think about a building as a system; and (3) use technology to make building(s) smarter and more transparent. The opportunity for existing buildings to become high-performance buildings demands accountability, with properly placed roles and responsibilities on every team member. Building

Box 1.2. *continued*

Renewables Last

Passive-building strategies reduce loads which results in active-building strategies that cost less and consume less energy. As a final step, renewables can be used to zero out remaining energy consumption and carbon emissions. At this point, on-site energy generation, photovoltaic arrays, geothermal well fields, and wind farms are more affordable due to their smaller size and lower first costs. In the future, replacement costs for the renewable solutions are naturally reduced, providing advantageous life-cycle costs for the final renewable solution. Building owners who believe that renewables are the silver bullet to energy efficiency and adopt them before adopting the first two steps of the Natural Order of Sustainability are discovering that the return on investment of a renewables-first energy strategy developed in isolation does not make financial sense when analyzed as first-costs or life-cycle costs.

Many believe that installing rooftop solar panels will resolve many of a building's energy sins. While they certainly help, the problem is that there is just not enough real estate on the roof of most buildings to handle the total building loads. Quality improvements to the envelope will last for 50 to 100 years. If limited dollars are available for a project, then putting the dollars into improvements that prevent the loss of heating and cooling energy makes more sense than adding more active equipment to mitigate the losses. When you look at the thermal image of a typical existing building on a 20-degree day and the building is blazing yellow or orange, the heat loss in the image may indicate a surface building temperature of 60 degrees. Doesn't it make more sense to prevent the loss of energy before installing another piece of equipment to generate more energy to make up for that loss? After all, the cheapest form of energy is, naturally, the energy never used.

---

owners who leave themselves room to bring new tools and ideas to the project team will have the greatest success. Goals and targets can always and should always be refined, but strategies should never substitute for building performance goals. When that happens, misalignment occurs, and frustration and confusion ensue. As an example of misalignment, think of your past project teams who installed bike racks at a building—bike racks that will likely never be used—solely to achieve a few extra LEED Sustainability Program points?

Establish Performance Goals

Trying to secure sustainability certification for a project that is misaligned with the project goals will lead to frustration as the project team struggles with competing priorities. If the goal is to reduce energy consumption or improve indoor air quality, then lead with goals using performance-based metrics, and let the sustainability certification plaques follow. By creating a data-driven, evidence-based environment, all voices have equal weight and all ideas are tested to determine the optimum solution. These foundational elements are driving the evolution of performance accountability in the built environment. Accountability is reached when building owners are called to defend the choices they made regarding the performance of their building(s). This accountability is an inflection point for owners and decision makers and it requires a holistic approach to building performance and sustainability.

Advocates for sustainability have their work cut out for them keeping project team partners focused on goals and performance accountability while aligning strategy. This is not a one-and-done approach to projects, but instead requires planning for short, medium, and long-term results. Creating opportunities for innovative ideas results from broad collaboration with stakeholders involved or impacted by the project. The process of collaboration may at times seem unnecessary to team members until they witness the benefits of aligning all team members to a common set of performance metrics and then bringing them together to consider and discuss the building as one holistic system.

Buildings Are Systems

Donella Meadows, author of *Thinking in Systems,* offers that "a system is an interconnected set of elements that is coherently organized in a way that achieves something. If you look at that definition closely for a minute, you can see that a system must consist of three kinds of things: elements, interconnections, and a function or purpose."[2]

Buildings can be complex systems interconnected in such a way that they produce their own pattern of behavior over time. If you change the performance of the building envelope, the mechanical systems are impacted. If you address lighting in the building, your final solution has a relationship between the windows and artificial lighting systems, which in turn impacts the heating and cooling systems. Simply addressing one element at a time can have significant unintended impacts on the building's function or purpose due to the complex interconnection of those elements.

Meadows articulates that we cannot impose our will on a system. But, by understanding the systems relating to buildings, we can discover how its properties and our goals and objectives can work together to bring forth something much better than could ever be produced by our will alone. When considering a retrofit project, these leverage points or "system triggers" can be used to consider broader, whole-building solutions. Because of the complexities of the building systems, it is difficult if not impossible to understand what impacts building performance. Fortunately, the technology now exists to model and simulate existing buildings to test those interrelationships. Using meters and sensors to improve and calibrate building modeling increases the accuracy of those simulation tests and reduces the chances of error. The same meters and sensors used to improve the model remain in place and become the feedback loops during operations, which serves to continue to refine building performance.

Without basic smart building infrastructure designed to provide feedback on the elements, interconnections, and a function or purpose, what we find is that buildings are nonlinear, turbulent, chaotic, and full of surprises. Yet most building owners and solution providers still attempt to improve the system by modifying the individual elements without respect to the system. From a systems-thinking perspective, this is the definition of insanity. As Meadows establishes, the dynamic behavior of a system cannot be understood or controlled just by tweaking elements of the system.

To see your building as a living, breathing system, it's essential that you find a partner with whom to develop an operational, calibrated, whole-building performance model. Using baseline data, as we will review in an upcoming chapter, you will get a first look at the building's overall performance. Innovative modeling software like IES VE or EnergyPlus give building owners insights they haven't had in the past.

Once you have a fully functioning model of your building, you can stop chasing symptoms (i.e., increasing utility bills or occupants' complaints about comfort levels) of root problems (i.e., leaky buildings, or poorly operating building controls). Building owners who simply address symptoms get short-term relief, but they must endure long-term frustration. Let's face it, opting for symptomatic interventions is tempting when it is the only obvious course of action. Because we don't always see the structures underlying symptomatic interventions, we focus on symptoms where the stress is greatest. Pressures to do something about occupant complaints are temporarily relieved. But relieving symptoms reduces the immediate need to determine the root causes. As long as root causes of building performance problems remain unaddressed, the building will continue to underperform. The most basic value of a model and its simulation capability is that owners and their teams are able to interrogate buildings in order to determine the root causes of performance issues.

Existing Buildings Must Become Smarter

When owners can see the whole-building solution and related performance, they typically want more data with which to develop more insights in order, ultimately, to develop more confidence in their next steps. However, it's easy to become overwhelmed by the magnitude of smart-building "solutions." This book does not endeavor to review all the potential ways people have discovered to make buildings smarter. Our bias is this: owners deserve the cheapest and easiest solutions that will enable them to visualize and control the performance of their building(s) based on the goals they set in planning.

If your building allows you to replace your network architecture, we suggest you first consider a building management system with "open integration" network and controls. Because the operation of buildings is so complex today, an early review of any renovation should start with a review of the building management system controls. If you can move from closed proprietary systems to an open-integration network approach, you will have a single platform for data management across your building(s). This one move will make anything you do for the remaining life of your building easier and cheaper.

Regardless of what you can do with your building management systems, the base components of a smart building infrastructure are an operational whole-building performance model, primary-source utility meters, indoor air-quality sensors and a JACE network controller. With a whole-building performance model and smart building infrastructure, you have the basic tools to begin making your old building perform like a new one.

The last piece of technology that is just beginning to surface in the market is an integrated building-performance dashboard. What separates this dashboard from others available on the market today is the integration of dynamic data across disparate systems including the simulation environment. The performance goals established and tested in the whole-building performance model are dropped onto a dashboard next to real-time trended performance data so that an owner can easily visualize how a building is performing against the project goals. It provides the context for performance from an investment point of view. Today, dashboards typically show some aspect of building performance over time, comparing historical data. This approach can only answer the question of how a building is performing against itself, historically. An integrated dashboard answers the questions, "Am I winning or losing?" and "Did I get what I paid for?"

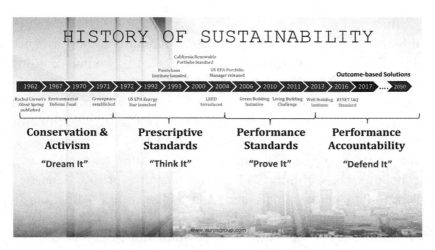

Figure 1-1. Longitudinal Look at Sustainability and the Built Environment.

Buildings and Sustainability

To fully appreciate the book's road ahead, it's important to understand the history of sustainability in the United States (see fig. 1-1). In the late 1960s and early '70s, sustainability was defined by conservation and activism. Sustainability advocates were few and their challenges were to try to turn the tide of years of waste and harm to the environment. Sustainable thinking and design standards for the built environment started with the American Institute of Architects (AIA). The AIA was the impetus for the creation of the United States Green Building Council (USGBC). In 1995, the USGBC partnered with the Natural Resources Defense Council (NRDC) to envision the first LEED green-building-rating system. LEED v1 was adopted by the USGBC in 1998 and served to provide prescriptive checklists of how to design a higher-quality building. The idea was: If you do this, you will get that.

LEED set the stage for thinking differently about designing and constructing better buildings. Around 2007, the prescriptive standards migrated to performance-based standards as witnessed with Living Building Challenge, Passive House, and other certification programs. The current state of sustainability is rapidly moving from performance-based standards to performance accountability as witnessed with LEED v4.0, Living Building Challenge, RESET Air, WELL Building, and other certification programs (see box 1-1 for more information on important certification standards). Federal, state, and local governments have added incentives and consequences through more-aggressive building policies, financial incentives for reduced energy consumption, building stretch codes, zoning regulations tied to building performance, and energy-disclosure ordinances in which building owners are required to disclose their energy consumption. There are many sustainability advocates who would argue that the efforts of governments and sustainability certification programs are not enough to offset climate change and change the course of the built environment. Nevertheless, there is no denying the macro-trend that our industry is rapidly heading toward performance advocacy, and any progress toward that goal will meaningfully impact climate-change and global-sustainability goals.[3]

As this book is going to print, New York City passed a building mandates law making it the first city in the world to require all large existing buildings of 25,000 sf or more to make efficiency upgrades that lower their energy usage and emissions, or face steep penalties. In 2015, buildings generated 67 percent of the entire city's greenhouse-gas emissions.

In fact, buildings are the primary source of greenhouse gases around the world (see fig. 1-2). In this context, they look at the concept of energy productivity and how effectively energy is used per unit of GDP. Here, residential and commercial buildings end up presenting about 34 percent of the opportunity to energy improve productivity. "When compared to other sectors, the buildings sector has the largest unrealized potential for cost-effective energy and emission savings."[4] In New York City, there are roughly 50,000 such buildings. This is a sign of legislation

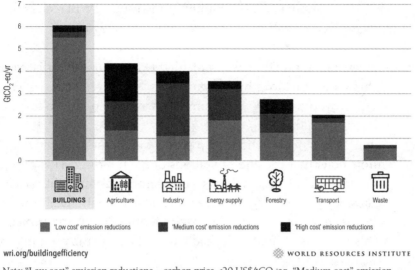

Note: "Low cost" emission reductions = carbon price <20 US$/tCO$_2$eq. "Medium cost" emission reductions = carbon price <50 US$/tCO$_2$eq. "High cost" emission reductions = carbon price <100 US$/tCO$_2$-eq. (Source: IPCC Fourth Assessment Report: Climate Change 2007: Synthesis Report.)

Figure 1-2. Economic Mitigation Potential by Sector, 2030 (used with permission from the World Resources Institute [WRI]).

to come all over the world. Winning building owners will find holistic, affordable solutions to meeting these performance goals; losing building owners will dramatically overspend on renewable solutions and suffer greatly when these "solutions" break.

Embedding Sustainability into Buildings and Culture

In the following sections, we summarize what we have learned from both research and practice about embedding sustainability into existing buildings and culture. We also draw information from analogous cultural-intervention efforts by sustainability certification programs, the indoor environmental-quality movement, and studies of high-performance buildings, and we attempt to draw parallels to embedding sustainability in decision-making.

The International WELL Building Institute (IWBI) classifies influencers in the built environment into four categories.

1. Green-Lighters: Those who greenlight sustainability initiatives. This group includes building owners, employers, developers, and investors, all of whom are ultimately responsible for the decision making.
2. Stakeholder Team: Those who participate in the strategic planning, construction, and operations of the building. This group includes architects, engineers, contractors, sustainability advocates, project managers (internal and external), facility and maintenance managers, and any practitioners on the project. This group is ultimately tasked with delivering the building to the occupants.
3. Other Audience: Those who use the building and are otherwise affected by the building. This group includes occupant and tenants, manufacturers, public health representatives, federal, state and local governments, and community.
4. Skeptics: Those who simply don't believe that it's possible to get old buildings to perform like new ones. This group typically includes people who are resistant to change and can be architects, engineers, owners, developers, etc. Even in the face of physical evidence and a boatload of examples of old buildings performing like new ones, they can't see the pathway to change their traditional design or construction process to take advantage of a powerful industry trend.

Internal selling of the potential of existing buildings is critical to the success of any sustainability initiative. Recent findings from MIT Sloan and the Boston Consulting group find that "sustainability-driven innovators do not treat sustainability as a stand-alone function detached from the business. They integrate their efforts into operations and planning."[5] Selling and making the business case for your project occurs at various levels: finance, legal, processes, data, systems, information flows, and the extent to which the participants invest in each other's systems. Integration of your project with the strategy of your organization involves parties working closely together to ensure the synchronization of their actions and decision-making processes. Internal understanding of the value your project can bring to an organization is important for collaborative information sharing and the resulting cross-functional and operational relationships necessary to enable the success of any project.

Whether or not you are successful with developing and maintaining a sustainability mindset and delivering building performance depends to a large extent on whether the people in the organization accept and believe in sustainability initiatives. In many cases, this means making sustainability part of the corporate culture.

Organizations affect how their members see issues, deal with problems, and identify what is important. People are influenced by organizational goals, structure, training, coworkers' attitudes, successes and failures, and a host of other aspects of organizational life. High-performance buildings such as those we discuss in this book can have large impacts on organizational culture, and different cultures may be more or less appropriate for a given set of goals for your building.

A few years ago, we joined a project team at the East Liberty Presbyterian Church (ELPC) in Pittsburgh, Pennsylvania. The church, located at Penn and Highland Avenues in Pittsburgh, was built in the 1930s in the Gothic style, a signature building of the renowned architect Ralph Adams Cram (see fig. 1-3). ELPC was planning their first capital campaign since the cathedral was built. The project goals included mechanical systems upgrades and building refurbishment, among other improvements. The culture of ELPC is steeped in social equity, inclusion, organizational stewardship, and environmental responsibility. ELPC was committed to undertaking a comprehensive renovation of the church building with the intention of becoming better stewards of our built and natural resources.

In recent decades, ELPC has progressively expanded its service to an ever-broader community as they seek to live out the dual missions of service and hospitality. The building has become a center of activity, drawing people from throughout the region not only for worship but also for education, economic opportunity, fellowship, and working together for justice and peace. ELPC is pleased to have the problem of accommodating such a full agenda of activities in the church building. It is a busy place, seven days a week.

Despite clear and overwhelming evidence of a culture dedicated to sustainability, ELPC struggled early on with its project goals. The first schematic design ended up with construction costs that were double the intended budget, annual operational expenses that were not affordable, and too many unknowns with performance goals. This reality, along with the leadership at ELPC adamantly advocating meeting its performance goals, forced the project team to reassess its traditional approach to design and construction.

ELPC invested in a whole-building performance model and incorporated important data for air-infiltration testing, digital metering, indoor air-quality sensors, inventory, and life-cycle assessment of all active systems. Using empirical evidence, ELPC created a comprehensive Owner's Performance Report to help align project team members to the project goals and targets.

The outcome of the modified approach to the project resulted in a final design that was within the project budget, met the goal of using 30–40 percent less energy than in pre-construction performance, and provided superior indoor air quality throughout the entire cathedral. ELPC completed construction at the end of 2018, so it is too early to confirm that the energy goals are being realized, but the current energy trends do meet the predicted performance targets. Lastly, in December of 2018, ELPC became the first cathedral in the world to achieve RESET Air Certification for Interiors.

ELPC is proud to tell everyone that its renovation is enabling them to increase their use of the building while limiting their use of energy. They are approaching the goal of sustainability in a comprehensive way, utilizing newly developed analytical tools to target facility improvements in a way that will be most effective

Figure 1-3. The East Liberty Presbyterian Church.

in conserving resources. Their installation of smart building infrastructure and an integrated performance dashboard provide their maintenance team the opportunity to level up, spending less time chasing after complaints and more time anticipating potential problems. The vast real-time data at ELPC, in the form of trended and targeted building performance metrics, has created a unique partnership with Carnegie Mellon University and Duquesne University. The first of its kind in the region, these two esteemed institutions can use real-time data from an existing iconic building for research on building performance. At eighty-four years old, East Liberty Presbyterian Church is one of the smartest buildings around.

The manner and degree to which your organization adopts building sustainability goals will very much depend on your organization's culture. For example, cultures that evolve over time in a Total Quality Management (TQM) or lean system emphasize waste and variance reduction, along with process standardization and discipline around process level routines. This cost-based approach to management can be used to showcase long-term cost reductions of investments in existing buildings. Such an approach may seem stifling to employees rewarded for creativity and radical innovations. In this way, lean initiatives can greatly affect the culture and work life of employees. Within innovative cultures, renovations and deep retrofits of existing buildings should be positioned as exciting and new endeavors (i.e., as a way to showcase innovative, green, high-performance materials and technology). Industry 4.0 spaces are now seeking to incorporate sustainability, innovation, and the reduction of wasted space and motion into existing buildings.[6] This way, innovative spaces encourage innovative people to want to come to work each day and develop services and products for customers who will be able to see this innovation not only in the organization's value proposition, but also in its built environment.

Knowing your audience begins with the development of a compelling story about the potential of your building or project idea. Each existing building contains within it several energy-savings projects around which you can craft a compelling story about the value of your project. Opportunities may also exist in transforming the building envelope, air tightness, insulation, roofing, windows and doors, lighting, mechanical systems, and water fixtures. Impacts beyond energy and water savings can be found in improving indoor air quality and human health, wellness, and productivity. Knowing your audience also means knowing the business opportunities for your building and proposed project. Building the business case for existing buildings involves identifying the benefits, risks, costs, technical solutions, timing, impacts on operations and maintenance, and alignment of organizational capability to deliver the desired outcomes. You want to demonstrate that "we can afford this" or, alternatively, that "we can't afford not to do this." Existing-building opportunities are found by identifying the root problems of the current situation and demonstrating the benefits of your vision for the future state of the building.

Leadership

A critical element of cultural change and having the ability to see the potential of existing buildings comes in the form of leadership. We can draw from two examples of leadership that helped create cultural change. The first involves a city policy initiative, and the second involves the ability of leadership to see opportunities for innovation and metrics to be part of learning for all stakeholders. The first example comes from the City of Pittsburgh, which, under the leadership of Mayor William Peduto, established a Building Benchmarking Ordinance in 2018. The ordinance requires building owners of nonresidential buildings over 50,000 square feet to present annual reports of their energy and water consumption to the City for benchmarking. Since 2018, owners of nonresidential buildings larger than 50,000 square feet are required to benchmark for the first time by June 1, 2018, and yearly thereafter. So how does this change culture?

The policy calls for a change in data collection and transparency. This transparency, in turn, is changing culture around how to measure and manage low-performing buildings. This is discussed further in chapter 2. The Pittsburgh ordinance starts with "benchmarking" a building by measuring its energy and water use and using that data to compare its performance over time, as well as to compare it to similar buildings. Benchmarking allows owners and occupants to understand their building's relative energy and water performance as well as wastage. They can use that information to make strategic decisions that will potentially save money and energy while improving comfort and health.

Building owners are required to use a secure online tool developed by the US Environmental Protection Agency (EPA), called the Energy Star Portfolio Manager, to track their energy and water use and submit their data to the City of Pittsburgh. This data informs building owners about a building's energy and water consumption, and it tracks progress year over year. Owners can use this data to assist in decision making for maintenance and upgrades.

Building Benchmarking Ordinance compliance data is available publicly. The City publishes annual reports summarizing the performance of Pittsburgh's large buildings portfolio; these reports visually illustrate performance on the City's online platform to show which buildings are participating, exempted, or eligible and nonparticipating, as well as buildings whose owners chose to participate voluntarily. The benefits from this ordinance (which are tracked by the City), along with change in regional culture regarding building data and transparency, are far-reaching and include improvements to the local economy, public health, and low-income housing, as well as a reduction of GHG emissions; the data is also a stepping stone toward achieving the city's 2030 climate change related goals.

Next, consider a recent project we completed for the Environmental Charter School in Pittsburgh. Their culture and pedagogy, fundamentally built on the environment, combine academic rigor with an "out the door" learning approach rooted in real-world problems. Students go outside to study places and spaces, and connect content to the environment and the community. When it came time to design a new middle school, they quickly saw the opportunity to embed their culture of sustainability into the building itself. Once they set highly aspirational performance goals for their building, they moved to complete the circle and established robust curricula that use the building to teach students how to monitor building performance and troubleshoot real-world building performance problems. This is the finest example we have seen of the full integration of culture, innovation, pedagogy, and the measurement of results. According to Nikole Sheaffer, the school's director of innovation and outreach,

Any project team is susceptible to conflict. Despite every team member's best intentions, the series of hand-offs present in a construction project often leaves members defensive and argumentative. Once performance metrics were established, we found our team more enthusiastic about collaborating on our new middle school. We noticed less emotion, less conflict, and more creativity. Specific goals made a noticeable difference, and it was such a breath of fresh air for all of us.

Typically, though, managers must find and address conflicts between organizational goals and existing cultural norms. In fact, preexisting cultural norms and standard operating procedures often form serious impediments to organizational change. If you start hearing that "this is the way we have always done things," it's a good indicator of resistance to change. Therefore, in environments of rapid change involving building standards and technology, decision makers must be ready to utilize the strengths and weaknesses of their organization's culture to then sell existing-building opportunities internally. These strengths and weaknesses are often difficult to identify. As one school administrator said,

Our board was used to traditional approaches to construction. It was difficult for them to understand that even though we weren't spending more on construction overall, we needed to spend a bit more up front in soft costs to get a dramatic improvement in whole-building performance and a reduction in long-term operating costs in the end.

Organizational culture plays a critical role in achieving sustainability goals. People within the organization must embrace and support the organization's view of sustainability for goals to be met. This is not always easy. There is disagreement and controversy surrounding some sustainability issues, like global warming, carbon neutrality, and occupant benefit. Leadership plays an important role in defining the culture and related sustainability goals for environmental performance and human productivity into cost savings and differentiation in the marketplace. For example, Herman Miller, a furniture company in Zeeland, Michigan, has had great success with sustainability.

One of the founders of Herman Miller believed strongly in corporate stewardship and responsibility. In large part, the company's commitment to sustainability stems from the values and corporate culture created by this founding leader. The role of culture doesn't just apply to companies. Take Pittsburgh, Pennsylvania, for example. In 2014, Pittsburgh was selected from over 1,000 cities to become part of 100 Resilient Cities (100RC), a program pioneered by the Rockefeller Foundation. Other cities in the 100RC include Los Angeles, Chicago, London, Paris, Rotterdam, and Bangkok. Because of the sustainability aspirations of the City of Pittsburgh, under the guidance of Mayor William Peduto, Pittsburgh is playing on the global sustainability stage alongside much larger cities. Culture is empowering and has a dramatic effect on goals and aspirations. If you start a project and you don't have organizational culture to support you, you may want to consider tackling a different project.

The reality of a project team is that there are multiple organizational cultures that will impact the team's effectiveness. The building owner's culture should reign supreme, but the cultures of the architectural firm, the engineering firm, and the firms of every specialist you bring onto a project will impact the way the sustainability goals are prioritized and achieved, from design through to operations. In many parts of this book, we emphasize the importance of carefully selecting project team members. As you can see from the details in this chapter, you should try to understand the organizational cultures of your partners. The attitudes, opinions, and beliefs at the top of these organizations, as they pertain to sustainability, are important considerations that will, in small or in large ways, affect whether or not your project will fully realize your building performance goals.

Conclusion

It is possible to make old buildings perform like new ones without paying a premium in construction costs. Using modern building science and innovations in technology, an existing building can become smarter, thus providing owners with pathways for confidently investing in energy-conservation measures and indoor environmental-quality strategies, in the proper sequence, at the right time.

Owners no longer need to feel overwhelmed or paralyzed when they think about what to do next with their building stock. With clear goals and an experienced team committed to building performance, it's appropriate for owners to expect a master plan of investment that will clearly articulate the steps to restore any building to its original intent at or below capital- and maintenance-budget expectations.

The balance of this book provides the tools, processes, and sequencing necessary to build owners' confidence in investing in existing buildings in order to reduce energy consumption and dramatically improve indoor environmental quality.

What we are experiencing right now in the construction industry is nothing short of a renaissance. Today is the most exciting time to work in the built environment. We all are fortunate to have a front-row seat to an epic transformation in building systems and performance.

The bottom line is this: if we can simultaneously reduce greenhouse gases (waste), conserve energy, and improve the overall health of existing buildings, we will have established the most impactful approach possible to extending the life of our planet and providing a healthier indoor environment for people—irrespective of location, outdoor air quality, and socio-economic position. Making old buildings perform like new is simply a matter of turning the traditional construction process on its head and using modern technology to shepherd investment such that every dollar has maximum impact on energy reduction and the improvement of indoor environmental quality. Any action taken on an existing building must be expected to improve the overall health and performance of that building, and an owner should be able to demand proof that their investment will lead to real results in operations.

Now that we have your attention, go to chapter 2 to learn how to get started on making your old building perform like a new one.

Project Development Homework

- ✔ Identify a building/project you know of that is in dire need of renewal. Ideally, the building serves essential purposes and could never be considered "disposable" to the community in which it exists.
- ✔ Start thinking about how and why you want this existing building to be a high-performance space.
- ✔ Think about how transforming the building aligns with the values of the owner. In narrative form, write out how you would articulate this option to the building's owners. You may have to make assumptions around the organization's values, but that's okay.
- ✔ Think about what the owner has told you is important about the building, and that will enable you to begin getting a sense of their basic values.
- ✔ Storyboard your narrative as a successful progression of your project/building through time to a future state of becoming a high-performance building.

Chapter 2:

Where Do I Start?

When it is obvious that the goals cannot be reached,
don't adjust the goals, adjust the action steps.

— Confucius

M OST PEOPLE ASPIRE TO DO the right thing. After working
in and around the construction industry for the last thirty years,
we can state emphatically that the vast majority of project team
members are looking to contribute their expertise to create the best
buildings possible. Building owners, given the choice, want the best
buildings that money can buy. Yet the industry struggles to understand how best to scale high-performance buildings and where to
start. The decisions building owners make that seem best in the
short-term show that they rarely consider how to allocate scarce
resources over longer periods of time. This misalignment creeps in
when stakeholders, who typically operate as cost centers, do not
trust new entrants into the decision-making process. Instead, they
look only at first costs, fail to utilize data analytics to understand
energy efficiency and indoor environmental quality, and miss the
long-term cost savings of an integrated approach.

Building owners who have successfully transitioned from a traditional process to an integrated one have done so by requiring continuous project alignment and realignment. They ask the right questions in early planning, maintain their convictions in their goals, and ultimately raise expectations for project results. Building owners who focus solely on first costs and limit their construction goals to building code compliance inadvertently force the project stakeholders to race to the bottom. Similarly, facility managers who simply focus on the symptoms of building performance have the same mediocre results. Operations folks have long told stories about installing "dummy" thermostats that are not connected to building controls as a means to slow or stop occupant complaints about thermal comfort and air quality!

Establish a Baseline

Just when setting any plan, buildings owners need to discover where they are before they can develop strategies for where they want to go. They must start by understanding how their building is currently performing. The information and data collected on building performance is necessary to establish a baseline for current utility consumption and indoor air quality. It is always best to collect at least 24–36 months of utility data in order to establish a baseline against which you can compare future building energy performance. The baseline assessment is used to set energy-efficiency improvement goals. Establishing a baseline helps decision makers understand how energy expenses contribute to operating costs and provides historical data as a context for future decisions and actions. It can also be used to identify high-performing buildings for replication of best practices and to prioritize poorly performing buildings for improved performance and retrofits. Further, baseline information is crucial to understanding social benefits. For example, indoor environmental quality and greenhouse-gas emissions are becoming lightning rods for many communities—especially low-income communities, which are disproportionately affected. The path to social equity requires community engagement, which is enabled when stakeholders have empirical data, performance information, and clear-cut goals.

Whole-building performance data drives the establishment of performance goals that help determine if the plan and the financial investment into the plan are reasonable. Reducing costs begins with understanding how utility expenses contribute to operating costs. Using Internet of Things–based technology solutions and real-time data, building managers are able to document energy-conservation efforts and compare them to the baseline in order to make the business case for their planned improvements (see chapter 4 for more on creating a business case). This enables building stakeholders to categorize utility use by fuel type, business unit, building, product lines, end use, and other factors to better understand and manage enterprise operations. They can use this information to establish thresholds for initiating retro-commissioning activities, setting building performance goals, and rewarding successful projects within an overall comprehensive plan.

A comprehensive plan should start by clearly defining both short-term ("first costs") and long-term ("operating and life-cycle") costs. First costs represent the construction costs to build new buildings and/or retrofit existing buildings. First costs are the building owners' budget lines available to invest in buildings. Using first costs and current building-code standards, building owners historically have assumed that project teams are aligned and that sustainable solutions can be simply "bolted on" to the current system. The problem with this picture is that current building code standards represent, at best, the worst performing buildings permitted by law. Further, when project budgets become constrained, the sustainability elements are typically the first things that are cut to realign to first costs.

We have inspected more buildings than we can count that utilized superficial energy conservation strategies like resetting the thermostats 1–2 degrees, turning off ventilators and economizers, closing window shades during the day while turning on light fixtures, and leaving windows open while heating and cooling. Without a comprehensive plan and a calibrated whole-building energy model, decisions to improve energy consumption and indoor environmental quality frequently have unintended negative consequences.

Many building owners and project teams are just now beginning to understand that it is possible to improve the performance of most buildings during design without increasing first costs. The general lack of understanding on this point can be attributed to a lack of goals beyond reducing first costs and meeting building code standards. Without declaring specific and measurable building performance goals, any problem that arises, like a delay of schedule, will result in most project teams simply reverting to traditional building processes and code based decisions. What if owners were able to utilize a different set of building performance goals that protect first costs and long-term operational costs?

Only when building owners recognize a holistic approach to the building cycle can they begin to influence the process toward better outcomes by setting very specific quantitative goals. When project teams focus on the fundamental goals of whole-building performance, four questions serve to align all stakeholders to specific performance metrics over the course of building projects, from conceptual design through construction and into building operations:

1. What is the baseline?
2. What are the whole-building performance goals?
3. What are the building costs to build or retrofit? (See chapter 4)
4. What are the building costs to operate annually? (See chapter 4)

This gets to the heart of what we see as a visualization of the owner's three most important steps (See fig. 2-1). Over the course of this book, we will use these as touchpoints for the planning and execution of existing-building projects.

Figure 2-1. Decision Matrix for Owners.

After baselining the building (discussed above), whole-building performance goals should receive the highest priority because they serve to balance building performance with both short-term and long-term costs.

What Are Whole-Building Performance Goals?

When establishing goals and targets, building owners need to consider the needs of building occupants and the intended uses of the building. After all, buildings are built to be used. If you cannot operate a building for its intended use, then the other two goals are irrelevant. What relationship do building owners want between the building and its occupants? If an organization believes that its people are its most important asset, then a building that doesn't provide a high level of indoor environmental quality would not be living up to that value. Building performance can be defined by the highly interrelated components of energy efficiency and indoor environmental quality.

According to Bako-Biro et al., "Building occupants have faster and more accurate responses to a cognitive function test at high ventilation rates."[1] The same study also showed that excess carbon dioxide (CO_2) can slow cognitive function, making occupants less attentive. (We will discuss how to set building performance goals in more detail in chapter 3 so that owners can determine what defines their unique balance between performance and costs.)

How these future savings are valued (and whether or not they include environmental and social performance) can make a big difference in the success of a project. A whole-building performance plan, in combination with data and analytics, can get more energy consumption and indoor environmental quality improvements out of existing technologies. For example, instead of repeatedly replacing HVAC units in kind time and time again, building owners can instead address the tightness of the building envelope. Integrated, value-maximizing solutions such as insulation, occupant access to operable windows, and real-time building-level dashboards can result in improved whole-building performance. This type of strategy can very easily result in a plan to install smaller HVAC units and mechanicals while simultaneously improving occupant comfort.

Energy efficiency is measured with a very simple, commonly used, whole-building metric. As we discussed earlier, energy-use intensity (EUI) is expressed as energy consumption per square foot per year. It is calculated by dividing the total energy consumed by the building in one year (measured in kBtu) by the total gross floor area of the building. The graph in figure 2-2 is based on research the Environmental Protection Agency (EPA) conducted on more than 100,000 buildings. It shows the median Energy Star Portfolio Manager (ESPM) source EUI vs. national EUI across thirteen building types. This key performance indicator has become the standard index used to compare building performance.

EUI is calculated as both site energy and source energy. Source energy represents the total amount of raw fuel that is required to operate the building (see fig. 2-3). It accounts for total energy use, incorporating all transmission, delivery, and production losses. Site

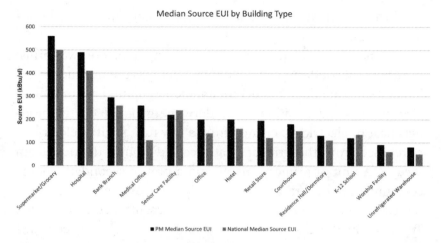

Figure 2-2. *Energy Use Intensity Performance (used with permission from the US Environmental Protection Agency summarizing benchmarked Energy Star Portfolio Manager data).*

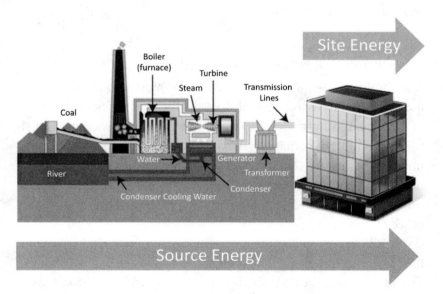

Figure 2-3. *Site and Source Energy (used with permission from the US Environmental Protection Agency Energy Star program).*

energy is the amount of heat and electricity consumed by a building within the building footprint.

Site energy may be delivered to a building in one of two forms: primary or secondary energy. Primary energy is the raw fuel that is burned to create heat and electricity, such as natural gas or fuel oil used in on-site generation. Secondary energy is the energy product (heat or electricity) created from a raw fuel, such as electricity purchased from the grid or heat received from a district steam system. A unit of primary energy and a unit of secondary energy consumed at the site are not directly comparable, because one represents a raw fuel while the other represents a converted fuel. Typically, in the case of existing buildings, most project teams pursuing energy-efficiency strategies focus on site EUI, as it is the easiest metric to understand and, outside of carbon dioxide, it is the second-least susceptible to manipulation. Conversations normally focus on the EUI resulting from site energy as the primary target for improvement.

The Importance of Indoor Environmental Quality

Any building owner considering deep renovation today must include indoor environmental quality (IEQ) as an essential factor of whole-building performance. IEQ is a major concern to businesses, building managers, tenants, and employees because it impacts the health, comfort, well-being, and productivity of building occupants. Most Americans spend up to 90 percent of their time indoors, and many spend most of their working hours in an office environment.[2]

An EPA report to Congress concluded that improved indoor air quality can result in higher productivity and fewer lost work days.[3] The EPA estimates that poor indoor air may cost the nation tens of billions of dollars each year in lost productivity and medical care. If you look for the largest line-item expense for an organization, you will most likely find it is the expense for employees.

Historically, buildings have been designed to create a uniform thermal environment that satisfies most occupants. ASHRAE Standard 55 stipulates only 80 percent of occupants need to be comfortable. The effects of indoor environmental-quality problems are often nonspecific symptoms rather than clearly defined illnesses. Symptoms commonly attributed to indoor environmental quality problems include headaches, fatigue, shortness of breath, sinus congestion, cough, sneezing, eye, nose, throat, and skin irritation; dizziness and nausea. By monitoring air quality on an ongoing basis, owners may know quickly and inexpensively if these physical symptoms are the result of poor indoor air quality.

Indoor air quality is not a simple, easily defined concept. It is a constantly changing interaction of complex factors that affect the types, levels, and importance of pollutants in indoor environments. These factors include the sources of pollutants or odors; design, maintenance, and operation of building ventilation systems; moisture and humidity; and occupant perceptions and susceptibilities.

Without data, perceptions of indoor environmental quality are as varied as the number of occupants in the space under consideration. Commonly received complaints of indoor environmental quality include odors, thermal comfort (too hot or too cold), air velocity and movement (too drafty or too stuffy), heat or glare from sunlight or from artificial lights (especially on monitor screens), physical aspects of the workplace (location, availability of natural light, aesthetics of office design, location and use of office equipment), ergonomics (height and location of computers, adjustability of keyboards and desk chairs), noise, and vibrations.

Some of these factors may be controlled by building management, such as maintenance of the HVAC system and the amount of outside air being mechanically brought into the building. Others are largely under the control of building tenants and occupants, such as the materials used in furnishing and products brought into the building. Some factors, like cleanliness and general housekeeping of the building, require the cooperation of both the building management and all the individuals who work or live in the building. For these reasons, indoor environmental quality is a shared responsibility.

Mitigating indoor air pollutants begins with good design, including a well-planned monitoring system. Fortunately, now there are tools, technologies, and processes that help building owners understand current and potential indoor air-quality performance. We advocate monitoring indoor air quality (IAQ) using commercial-grade quality monitors that can be easily and reliably recalibrated. There are many multi-parameter monitors available to measure and monitor a range of air-quality parameters, including temperature, relative humidity, carbon dioxide, total volatile organic compounds, and particulate matter measured in 2.5 and 10 microns.

Controlling indoor air quality involves integrating three main strategies. First, manage the sources of pollutants, either by removing them from the building or isolating them from people through physical barriers or air pressure relationships, or by controlling the timing of their use. Second, dilute pollutants and remove them from the building through ventilation. Third, use filtration to clean the air of pollutants.

Good indoor environmental quality management practices can make a big difference. It is also important to remember that any building, no matter how well operated, may experience periods of unacceptable indoor air quality due to equipment breakdown, inadequate maintenance, or, in some cases, the actions of building occupants.

Using indoor environmental quality sensors to identify real-time thresholds for key performance indicators of environmental conditions places building managers and occupants in a good position to maintain comfortable and healthy building environments. Success depends on cooperative planning, goals, and actions taken by building management and occupants to improve and maintain indoor environmental quality.

As an alternative to using sensors, if you ever want to know how buildings are performing, just ask the occupants. Soliciting feedback about environmental conditions from the people who spend much of their time there is logical, but often overlooked. Most building owners resist asking occupants—for numerous reasons. However, we have witnessed a transformation in thinking when occupants are solicited for their input regarding building performance.

Most indoor environmental quality investigations begin in response to a complaint from one or more building occupants. Indoor environmental-quality complaints can affect entire buildings or be limited to areas as small as an individual work station. The goal of the investigation is to resolve the complaint without causing other problems. In many buildings, thermal comfort is a never-ending battle for facility managers. Remember that it is imperative to identify the root causes of thermal dissatisfaction that are unique to your facility. Only then can you take the steps necessary to address them. Facility managers or engineers often overcompensate by adjusting thermostats ineffectively, leading to possible energy waste.

Problems vs. Symptoms

Most building owners and facility managers typically respond to symptoms of underlying problems. An uncomfortable occupant is a symptom, not a root cause. What we know from the 1970s total quality movement is that real improvements in systems come from addressing and resolving root problems, not reacting to symptoms of the problem. Wasteful buildings with changing thermal comfort levels, high CO_2 levels, or constantly running HVAC systems are indications of a problem, not the root cause. We can think about it this way: waste is anything that does not add value to a product or service, and waste can be discovered by following the trail of its symptoms. To help differentiate between symptoms and problems, ask the question "Why?" By using a total quality management (TQM) technique by asking the question "Why" up to five times, you will typically reach a sufficient level of understanding of the root cause of a problem. Why are occupants uncomfortable? When you get this response such as "It's too hot," then ask, "Why is it too hot?" Each "why" that you ask can take you another step closer to the real, underlying problem.

Addressing root causes through building materials and mechanical systems is often dismissed as too expensive, yet the unintended costs from inefficiencies in buildings can easily outweigh first costs. Waste in the form of greenhouse gas emissions from high energy consumption has additional costs to track and manage, and thus can be recognized as a building inefficiency. Simply stated, high energy use, high water use, poor indoor environmental quality, and greenhouse gas emissions from a building are all symptoms. Symptoms tell us that something has gone wrong, but they do not explain *what* has gone wrong. Symptoms may reveal the perceived magnitude of the problems, but they do not reveal the factors that have contributed the most to the problems. Stop for a moment and think about the implications of placing a dollar value on these inefficiencies, such as valuing the social cost of each metric ton of carbon dioxide equivalent (CO_2e) at $40 per ton of carbon emissions. How will this change owner and manager decisions as we look for new forms of efficiency and effectiveness within existing buildings?

You should diagnose and measure the indoor environmental quality in an objective and holistic manner using empirical evidence as well as guidelines such as those developed by trade associations.

Conclusion

The recent and rapid advance in building science means that most buildings in the United States over twenty years old have building envelopes that have high air leakage rates and require owners to increase the flow of heating and cooling to maintain thermal comfort and fresh air flow. Very few have optimized envelope design (see chapter 5), nor the optimal mechanical, electrical, and plumbing systems to balance enhanced envelope performance with right-sized active systems. It is easy to hypothesize that most mold and mildew conditions facing commercial buildings designed decades ago (i.e., K–12 schools, universities, hospitals, residential towers, and so on) are the result of poorly constructed building envelopes. Optimized envelope design, can make a significant difference. According to the Passive House Institute, Passive House buildings around the world are delivering energy savings up to 90 percent over traditional, code-based buildings, while creating a controllable environment for high indoor air quality.

The challenge for most owners of existing buildings is to thoughtfully and economically fund a long list of deferred maintenance tasks and systems at or near the end of their life cycle that would meet the minimum threshold for restoring the building to its intended design and purpose. Take, for example, condominium associations. Across the world, condominium associations try mightily to avoid having their buildings become "Zombie Condos." Condominium boards struggle with the cycle of aging infrastructure, increasing dues and assessments, and increasing numbers of vacancies. The fundamentals of condo ownership, where neighbors become business partners, is typically a game of whack-a-mole in which each maintenance issue is viewed as a discrete event, and each repair or replacement is put on a list where the priorities are reshuffled depending on the opinions and maintenance tolerances of the condominium owners and business partners. This confluence of issues causes a disturbing trend whereby residential towers fall into terrible disrepair, risking eventual building condemnation.

We believe that building science and recent innovations in evidence-based performance make now the right time for owners to consider a holistic approach to restoring existing buildings to their original design intent. We offer the following approach:

1. Financial Comparison: Take a holistic rather than incremental approach to building restoration.
2. Risk Comparison: Take a holistic rather than incremental approach to building restoration.
3. Façade Strategies: Take a holistic rather than incremental approach to building restoration.
4. Owner Disruption Narrative: Review strategies to reduce disruption to owners of a holistic building restoration.
5. Funding Strategies: Assess economic opportunities enabled by a holistic building restoration.

When building owners consider a strategic plan to address building performance, they would be wise to take a step back and understand the complexities of interrelated systems. To holistically improve building performance, consider the building as a living, breathing, dynamic system that is changing in every moment, like the human body. A thorough understanding of all the factors that interact to create indoor air-quality problems can help to avoid unintended consequences. It is specifically due to this complexity that whole-building modeling is necessary in order to test and weigh energy-conservation measures and bundles of measures. Calibrating the whole-building model with existing-building performance data validates the simulations from the model to make better investment decisions. Interestingly, the same smart building infrastructure used to create the baseline performance of the building is used to validate the model (see chapter 7). These steps will be discussed in detail in the following chapters.

Project Development Homework
   ✔ Consider the building you identified at the end of chapter 1. What would you say are the greatest opportunities for that building if it were able to be transformed into a high-performance building?

✔ What is the current performance of the existing building in terms of empirical data? This baseline information is necessary to begin planning.

✔ How might the owners of the building benefit if that building's performance were to become best-in-its-class?

✔ If you were the owner of that building, what would help you gain the confidence necessary to make the proper investment in the building?

Chapter 3:

The Importance of a Project Plan: Every Building Needs One

A goal without a plan is just a wish.
— Antoine de Saint-Exupéry

IN MOST CASES, THE EASIEST PART of the process of transforming an existing building is to find all the ways the building is performing below expectations. Whether it's because of inefficient use of energy or poor indoor environmental quality, buildings built even just ten years ago are underperforming and failing to meet modern occupant expectations. To begin laying out the business case for renovation, it's important to consider all the areas of occupant dissatisfaction. Building managers might consider a subjective survey of building occupants, as described in chapter 2, or they might establish a building-improvement team to be responsible for developing areas of potential improvement. Either way, the objective is to develop a prioritized list of improvements that would be most transformative if executed.

Once decision makers have developed an informal list of priorities, they are ready to conduct a formal sustainability charrette.

The purpose of a charrette is to bring together a broad group of committed stakeholders to build consensus around key aspirational sustainability goals. It's important to have a skilled facilitator conduct the charrette.[1] A charrette is an intense, project-focused or organization-focused approach to creating momentum around a project, and as such, it requires skilled facilitation. In many ways, the charrette is a process of discovery to identify the culture of an organization, as discussed in chapter 1, to ensure the alignment of project goals. Every area of the country has skilled facilitators for sustainability charrettes, but you should take care to hire a facilitator who is not "pre-wired" to a specific sustainability solution or strategy based on a specific sustainability certification program. Most sustainability facilitators have credentials in specific sustainability certification program standards. It is therefore important to tap a facilitator without obvious bias. That way, you can be sure the facilitator isn't guiding your project according to the credentials for which they are certified. Equally important is that your facilitator starts the process of aligning all participants to the organization's vision and values. This step roots all sustainability work in the elements most aligned with the culture and core principles of the organization. Taking this step gives you the greatest chance of getting top-level support to implement the sustainability goals.

To get a good sense of how a sustainability charrette is conducted, we refer clients to a National Renewable Energy Laboratory (NREL) publication called *A Handbook for Planning and Conducting Charrettes for High-Performance Projects* (2003) by Gail Lindsey, Joel Ann Todd, and Sheila J. Hayter.[2] While every facilitator has his or her own approach, this guide will give you a general sense of the preparation required to conduct an effective sustainability charrette.

Post–sustainability charrette, your facilitator will provide you with a report outlining the agreed-upon goals and targets for the project. In most cases, this report becomes the foundation for the

Owner's Project Requirements (OPR) report. The OPR is a living document intended to be updated periodically as building owner's make project decisions to ensure stakeholder alignment during the project's life cycle. Typical goals that become the foundation of the OPR are as follows:

- lower energy bills
- increased overall thermal comfort
- less indoor humidity in the summer
- more windows for natural lighting
- increased use of currently unusable spaces
- higher indoor air quality (less dust and so on)
- public and transparent access to building meters and sensor-trended data
- pathway to deliver the Architecture 2030 Challenge goals
- more independent user control over occupant thermal comfort.

Your report may mention strategies or standards that the sustainability charrette team discussed during the charrette process. Types of strategies that are mentioned in sustainability charrette reports include but are not limited to

- specific sustainability-program certifications;
- passive "first" strategies using Passive House principles;
- active "second" strategies to address thermal and environmental comfort using decoupled systems and/or mixed-mode strategies;
- water strategies for reducing and reusing storm-water runoff and reducing the use of potable water;
- material strategies using Living Building Challenge strategies; and
- renewable "last" strategies to offset energy generation with on-site energy generation.

As a facilitator team, we typically refrain from trying to direct clients toward certain pathways or strategies to achieve their goals; rather, we prefer to guide teams toward the development of their own goals and targets in the form of metrics. We like to ask how stakeholders will assess "success" at project completion. Or, when in doubt, we ask them, "What does winning look like?" If a group is defining success as securing a standard certification, then we include that as a sustainability goal. If, however, the team is looking for broadly improved occupant benefits, we recommend that they defer commitment to a sustainability standard until the goals, targets, and basic strategies are fully vetted. Otherwise, the project team may lose focus on the big picture while committing vast resources toward obtaining the sustainability standard certification. This is a bit like the tail wagging the dog.

Once the sustainability charrette report is distributed, it is time to begin creating options for setting the goals and targets of the project.

Owner's Project Requirements

An Owner's Project Requirements (OPR) report may be the most important document developed at the beginning of any project. We say it may be the most important because project success depends on what is included in the OPR. An OPR that is mostly "canned" copy or that tends to focus on ethereal wish lists will be celebrated for a day or two and then promptly forgotten and pushed aside because it cannot be acted upon. On the other hand, a working OPR becomes the written guide for a project that serves to align project teams to goals based on specific metrics. A working OPR is updated throughout the project anytime an owner chooses to modify goals or expectations. It is used to align team members, from design to operations, with the owner's expectations for the building's performance during operations.

Goals articulated during a charrette set the foundation for the

OPR, but the OPR must seek granularity if it is to be fully functional. For example, it's fine to discover during a charrette that owners are more interested in reaching ultra-high indoor air quality than they are in reaching zero energy consumption. However, that must be translated into specific metrics. What level of carbon dioxide or particulate matter 2.5 (particles less than 2.5 microns in diameter) is an owner willing to accept? Once owners reach this level of granularity in their expectations, it's time to consult sustainability certification programs for guidance on the articulation of the metrics in the OPR.

Whether you are pursuing ultra-low energy and water consumption, ultra-high indoor air and environmental quality, or both, certification program standards offer strategies and contexts to land on the right performance metrics for a project. Figure 3-1 compares some of the most established sustainability certification programs by performance metrics.

Remember, the key at this point is to land on the goals and targets based on specific metrics. We do not recommend committing to a sustainability certification program or standard until the project team first identifies its internal metrics.

There are two distinct decisions to eventually make on building performance:

1. Which metrics (not sustainability certification program standards) will the project team align to in order to meet them through the execution of the project? This defines the expectations for building performance during operations. The most successful projects reach this alignment to metrics very early in the conceptual phase of a project.
2. Is the owner completely set on pursuing a specific sustainability certification? The distinction here between applying certification standards verses pursuing the certification is important. To get lower energy consumption or higher indoor environmental quality, teams can use the strategies embedded in many

AUROS GROUP Targets	Energy Site EUI (kBtu/sf/yr)	Water (gallons)	Indoor Air Quality				
Sustainability Program Certification, Standard Performance Targets			PM2.5 (µg/m3)	TVOC (µg/m3)	CO_2 (ppm)	Relative Humidity (%)	Temperature (°F)
INTERNATIONAL LIVING FUTURE INSTITUTE LBC 3.1 — New and Existing Building Targets	Zero Energy *Post-Renewables*	Net Positive Water *Post-Renewables*	None	< 500 Red List	None	Measured & Monitored to Meet Project-Based Goals	
INTERNATIONAL WELL BUILDING INSTITUTE WELL v2 — New and Existing Building Targets	None	Water Quality Monitoring	< 15	< 500	< 800	> 30 to < 50 Measured & Monitored to Meet Project-Based Goals	
Passive House Institute — New Building Targets / EnerPhit Targets	14.0 20.0 *Pre-Renewables*	None	None	None	None	Measured & Monitored to Meet Project-Based Goals	
PHIUS Passive House Institute US PHIUS+ 2018 — New and Existing Building Targets	14.0 *Pre-Renewables*	None	None	None	None	Measured & Monitored to Meet Project-Based Goals	
RESET Version 2.0 — High Performance Targets / Minimum Targets	None	None	< 12 < 35	< 400 < 500	< 600 < 1,000	Measured & Monitored to Meet Project-Based Goals	
LEED v4 — Certified / Silver / Gold / Platinum	None	LEED v4 Minimum of 5% energy efficiency improvement over ASHRAE 90.1-2010	None	< 500	None	Measured & Monitored to Meet Project-Based Goals	

Revised 01/01/2019   © 2019 by AUROS Group

Figure 3-1. Sustainability Program Certifications and Standard Performance Targets.

sustainability certification programs without ultimately pursuing the standard's certification. It is very important to know early in a project whether an owner views "success" as meeting a specific certification's metrics, or hanging a plaque, or both.

The biggest mistake we see project teams make is landing on a certification standard before they fully understand the impacts (cost and performance) on their specific project caused by delivering that standard. By focusing on "which metrics constitute success in operations," owners are able, before they spend money on construction, to use data and technology to know how easily or cost-effectively they can reach their performance goals. Only at that time does it make any sense to land on the pursuit of specific sustainability certification programs. This approach will save project teams time, money, and frustration in the design phase of the project.

Upon completion of the discovery charrette and creation of the OPR, wherein building owners identify their goals and basic strategies for the building, it's time to short-list the various certification standards for consideration. (See box 1-1 in chapter 1 for a brief description of each program and appendix 2 for related resources.) For example, if an owner prioritizes the reduction of energy as the exclusive priority for a building, the WELL Building standard may not be the best standard to pursue, because WELL Building focuses more on occupant benefits achieved through the implementation of a healthy building environment. It is important to mention that it is not necessary to identify any specific building certification standard(s) during conceptual design or design development. In fact, many thought leaders in sustainability are using evidence-based design to align project teams with specific project goals.

Take the Rocky Mountain Institute's (RMI) Innovation Center, for example. RMI identified the metrics it intended to achieve in terms of energy-use intensity (19 EUI) and indoor environmental quality (expanded thermal comfort of 68–82 degrees Fahrenheit).[3] RMI's goals also contained aspirations for "a beautiful structure focused on community outreach and occupant experience." Once RMI created the path and alignment to achieving their goals, only then did they consider which standards they would pursue. In the case of RMI, they were able to achieve Passive House and LEED platinum certifications, while also meeting the Architecture 2030 Challenge goal of 70 percent energy reduction based on efficiency alone (prior to accounting for on-site energy generation).[4]

We are noticing a trend wherein owners are pursuing specific attributes of building sustainability certification programs without pursuing full certification. As practitioners of sustainability principles, we should be reasonably indifferent as to motive or intent. However, we would caution building owners to carefully manage this approach. For example, arbitrarily adding insulation and air barriers to the envelope may cause envelope-assembly durability issues. Also, right-sizing mechanical equipment without rigorously managing high-performance envelopes may result in the systems' inability to maintain thermal comfort. Mitigation of these matters can be managed with professionals experienced in whole-building performance modeling and commissioning, but it cannot be taken for granted. With an inexperienced team, building owners may be best served by pursuing full certification with those standards that align with their goals and targets. That way, the standard holds every team member accountable for their contributions. If every new building or major retrofit considers building performance standards or simply focuses on certain aspects of building performance, we would be off to a good start. The increased awareness of building owners, developers, and project teams of the importance of spending as much time on how the building will perform as they do on architectural programming requirements of the building will singularly propel existing buildings toward their high-performance potential.

Unfortunately, many building owners view their decision as a matter of energy consumption versus indoor environmental quality. We can state unequivocally that you do not have to choose one or the other! As we saw in the RMI example above, it's possible to drive toward ultra-low energy consumption and prioritize exceptional indoor air and environmental quality—without sacrificing first costs or long-term operating costs. Some project team members may offer skepticism based on the historical challenges of getting the envelope tight enough to reach low-energy goals while maintaining enough ventilation to ensure the benefits of high indoor air quality. But the simple fact is that there is too much evidence today to show that high-performance buildings are being retrofitted and built new without a premium in construction costs.

Today, building science and whole-building performance modeling, along with data analytics and smart building infrastructure, provide practitioners with the tools they need to ensure that a building reaches both ultra-low energy consumption and ultra-high indoor air and environmental quality. In fact, with the growing body of empirical evidence today showing the relationship between buildings and their occupants' own health and personal performance, an owner would be very foolish to discount the importance of indoor air and environmental quality. In subsequent chapters, we will share some of the many reasons for, and the financial benefits of, pursuing improvement in indoor air and environmental quality.

We offer the following tips to increase the odds of achieving building performance goals:

- Communicate goals transparently to the entire team simply and often. Create a one-page Owner's Project Requirements (OPR) document showing the goals and targets, using metrics as defined by the owner's stakeholder team.
- Periodically update and share the OPR to remind each team member of his or her obligations to the achievement of the goals.
- Each time the goals are communicated to project team members, ask for questions or concerns about the goals. Give team members the time and space to comment, complain, or clarify how they are feeling.
- Detail the commissioning and measurement and verification specifications early in the planning process. The project teams will benefit from knowing early how the owner will test, measure, and monitor building performance.
- Document the direct and indirect benefits to the owner for achievement of the goals as defined in the OPR.
- Routinely calculate the first costs and long-term operating costs of achieving the goals as stated in the OPR. There should never be a time or a decision point when the owner doesn't see the impact of any proposed change in building performance goals, first costs, and long-term operating costs.
- Remind the team of the steps it has taken to get to each defining moment in the project. By sharing challenges and successes, teams remain engaged and committed to one another and to meeting the upcoming challenges.

Finding your way through discovery charrettes and the goal setting process isn't straightforward. We often say that this is a very messy process. Doing it well means listening to the building owner, building occupants, stakeholders, and community, and giving them all the time and data to change their minds throughout the conceptual design process. As you navigate these stakeholders through the myriad of options, they will gain clarity in terms of what moves and motivates them. Once a team gets through the entire process on a project, the discovery charrette and goal setting parts of the process only get easier on each subsequent project.

Within a few miles of each other in Pittsburgh, Pennsylvania, three existing buildings are undergoing some level of renovation, representing three very different sets of needs and expectations as they relate to sustainability. (See chapter 8 for several case studies of project planning and execution that also integrate sustainability.)

First, there is a gothic cathedral that was built in 1935 and is the cornerstone of the local community. Its stakeholders' goals were to reduce energy consumption and significantly improve their indoor air quality. They followed Passive House building science and were recently named the first cathedral in the world to be certified in the RESET Air indoor air-quality standard. Stefani Danes, Adjunct Professor of Architecture at Carnegie Mellon University and Elder of East Liberty Presbyterian Church, explained their process:

> Once we defined our mission for how we wanted our historic church to be used, we set performance goals for reducing energy [use], improving indoor air quality, and reducing long-term operating costs. Our expert project team was able to collaborate on a whole-building performance model that simulated how our building would perform before we spent construction dollars.

Second, there is a charter school whose entire curriculum is oriented around the environment and is in the middle of renovating an abandoned middle school. Its decision makers decided to pursue very aggressive sustainability goals in the form of WELL Building and RESET Air. For their school, they felt that their investment was best served by creating an environment where students and staff would be the healthiest and the most productive.

Third, there is a university with a long-term renovation plan under way to improve the performance of a fifty-year-old academic building that houses the university's MBA program. While its stakeholders are interested in lower operating costs and higher indoor air quality, their overarching goal is to make the building part of their curriculum, providing hands-on education and research opportunities for MBA candidates. Essentially, they want the building to reflect the culture of the program. As such, they need data. Therefore, they prioritized their investments in tools and technology to show the building's real-time and historical performance data. Investments in whole-building performance modeling, along with meters, sensors, weather and outdoor climate stations, and an integrated dashboard, ensures that the building has the rigorous data needed to support future decision making as well as research and development. In this case, there are no standards or certifications under consideration at this stage of the renovation.

These three examples demonstrate that no two buildings should be expected to have the same building performance goals. Until owners fully form their expectations for the performance of their buildings, every step forward, as awkward or messy as it feels, demonstrates major progress.

Insights and Metrics for Your Building's Plan
Some of the best learning and insights come from the discovery and charrette process. Honest and open engagement in this process, along with a clear articulation of expectations and goals, become

performance metrics for the eventual plan and organization. While we cannot list or summarize all relevant metrics for a given project, we can share an example of an OPR we use on projects (see the sample owner's project requirements in box 3-1, below). This example highlights the critical metrics that we typically identify during our discovery and charrette process.

Identifying passive solutions first, active solutions second, and then looking for integrated solutions with the use of on-site renewable-energy generation is a new way of thinking about existing-building projects.

Conclusion

The essential components of any good plan for an existing building begins with simply identifying the energy and indoor air and environmental qualities in terms of metrics. This simple first step unlocks all the next steps necessary to complete the plan. Depending upon your goals and targets, you may not even need a plan. But if you are looking to improve the energy efficiency of your existing building or you are looking to improve the indoor air and environmental conditions to the benefit of the occupants, then the next component of the plan will be to identify the most impactful areas of improvement in the performance of your building. To complete the plan, you will have to finalize the construction budget for the improvements and establish the annual operating costs for the building. At this point, you are ready to unleash your project team to collaborate on the design and construction strategies to meet the plan. You are also ready to continue on to chapter 4 with the confidence that comes from knowing where you are and where you are going.

In this chapter we highlighted the opportunities and risks associated with setting building performance goals. The old construction adage, "Measure Twice, Cut Once," should be heeded when it comes to setting goals. Done well, this process aligns building owners' and other stakeholders' expectations for building performance

Box 3-1. Owner's Project Requirements

Duquesne University, Rockwell Hall	
Building Owner	Rockwell Hall
Building Address	Forbes Avenue, Pittsburgh, PA
Building Size	165,945 GSF, 13 stories
Building Type	College/University Campus Level
Construction Budget	$ TBD
Annual Site Energy-Use Intensity (EUI)	(tbd) kBtu/sf/yr
CBECS Site EUI	(tbd) kBtu/sf/yr
Annual Utility Costs	$(tbd) ($(tbd)/sf)
Owner Sustainability Director	(tbd)
AUROS360™ Advisor	(tbd)

Targets and goals

Sustainability Program Goals	LEED, RESET Air Certification
Energy	
Site EUI	20–30 kBtu/sf/yr, meet The 2030 Challenge (Year 2030)
Building Enclosure	
Thermal Envelope	$R \geq 38.5$ hr. ft^2 F/BTU, $U \leq 0.026$ BTU/hr. ft^2 F
Windows Installed	$U \leq 0.15$ BTU/hr. ft^2 F
Airtightness	≤ 0.6 ACH$_{50}$
Indoor Air Quality	
Temperature	Meet ASHRAE Standard 55-2013 Section 5.3 Standard Comfort Zone Compliance and 5.4 Adaptive Comfort Model
Humidity	Between 30% and 50%
Carbon Dioxide (CO2)	< 600 ppm
Carbon Monoxide (CO)	< 9.0 ppm
Formaldehyde	< 27 ppb (< 0.027 ppm)
Ozone (O3)	< 51 ppb (< 0.051 ppm)
Particulate Matter 2.5 (PM2.5)	< 15 µg/m^3
Particulate Matter 10 (PM10)	< 50 µg/m^3
Radon	< 0.148 Bq/L (< 4 pCi/L)
Total Volatile Organic Compound	< 0.4 mg/m^3 (< 400 µg/m^3)
Ventilation Rate	30% more outdoor air than required ASHRAE 62.1-2013
Sound	
Office and Classroom Spaces	< 40 dBA
Exterior Noise Intrusion	< 50 dBA
Reverberation Time (RT60)	0.5 seconds
Light	
Task Light at Workplane	300–500 lux
Ambient Light	215 lux

(Comply with WELL Light Feature 62 Daylight Modeling for spatial daylight autonomy & annual sunlight exposure.)

and total costs (i.e., first costs, and long-term operating costs). If not done at all, though, this lacuna in goal setting sets a project up to deliver very little. Projects that do not employ Owners Project Requirements always tend to drift toward the lowest common denominators of building codes and decisions based only on first costs. Team members unknowingly work to avoid conflict, which, in the end, puts building costs and performance at risk.

The breadth of potential sustainability programs and certifications means that there is a solution for every building, but owners and their representatives should first establish their own internal goals and targets before selecting a sustainability certification program, a prioritization that will ensure proper alignment. Teams should spend as much time as possible on the specific performance metrics before moving forward.

Project Development Homework

- ✔ Using the building/project you have identified so far, what specific goals would you set relative to energy performance, indoor air and environmental quality, first costs, and/or long-term operating costs?
- ✔ Can you break down whole-building goals or project goals by area of function of the building?
- ✔ If you can achieve those goals, who benefits from each goal?

Chapter 4:

Can I Afford This?

Only win–win companies will survive, but that does not mean that all win–win ideas will be successful. Managers need a methodology for discovering solutions that yield the greatest benefits.

— Johan Piet[1]

S O YOU HAVE AN IDEA OF HOW you want to improve an existing building, and you have found the people in your organization who will support that mission. Now what? How do you get approval of your ideas to make them a reality? Building your case through an evidence-based approach to the project will show your team an integrated path to making your old building perform like a new one. Try to keep a couple of things in mind. First, the potential value from improving an existing building is far-reaching, and, second, you are not alone in your search for solutions.

To ensure your projects and ideas get the visibility they need, consider showing not only the costs and benefits but also the sustainability and value-maximization opportunities of your existing building's vision. After tackling the steps outlined in the first three chapters of this book, you are now ready to calculate the benefits of the proposed building changes and develop a plan to communicate those benefits in a manner that aligns with your enterprise's goals.

By knowing the pitfalls to avoid, you will exceed expectations for understanding how the project will affect financial performance (profit and loss statements) such that others will understand how to evaluate the costs and benefits you propose. Any time you want to ask other stakeholders to invest resources, especially if this means going beyond an existing budget for new projects, you will need to show payback in the form of improved performance.

Build the Business Case for Existing Buildings: Save Money, Improve Human Health and Productivity, Reduce Environmental Impacts

According to Campbell et al. and information from the Harvard Business Review, older buildings are one of the biggest sustainability challenges for cities. With a median age of thirty-two years, what would the US built environment look like if 87 billion square feet of building space and 5.6 million buildings invested in whole-building solutions? How would the world improve if every owner of every existing building took our approach? The impact could be, as my daughter likes to say, *monstrous* (very big!). While this information about the total number of buildings worldwide is difficult to know for certain, we found one study that estimated there are 1.7 billion buildings. Ed Mazria of Architecture 2030 summed up this opportunity by saying that in most major American cities, like New York City and Seattle, less than 3 percent of the existing buildings produce almost 50 percent of the greenhouse gases.[2] We can quickly see that turning our attention to optimizing the performance of existing buildings has the potential to rapidly accelerate the realization of climate-change-mitigation goals around the world. It's really the perfect value proposition: to put the world's future first by leveraging the embodied energy of existing buildings, employing modern, Passive Building science, and thus creating a new legacy for existing buildings. We state the goal in the following way: ultra-low energy consumption and ultra-high indoor air quality that provides all people a healthy environment in which they can do their best work.

The business case for change includes detailing how green buildings save 30–50 percent of energy consumption and 40 percent of water usage, while they produce 39 percent less carbon emissions and 70 percent less solid waste. The average cost premium is 1–2 percent for a payback period of between 12 and 24 months and an average payback of 20 percent over the lifetime of a building.[3] There are years of evidence from the Pennsylvania Housing Finance Agency that lead to the conclusion that large multifamily Passive House–certified projects can be completed at no additional cost as compared to conventional construction. This local evidence is further supported by a major competition in Brussels, Belgium, over an eight-year period from 2006 to 2014. City administrators there found that the average cost increase for a low-energy, Passive House project was 1 percent overall. They further found that for larger, non-domestic projects, it was actually cheaper to build a low-energy Passive Building than a traditionally constructed building.

In terms of the well-being of building occupants, we know from empirical evidence that access to outside views helps with mental functioning and memory, enabling faster call-center processing and shorter stays in hospitals. Daylighting helps students achieve higher test scores and learn faster, helps workers be more productive, and increases retail sales. Productivity increases by 23 percent with better lighting, 11 percent with improved ventilation, and 3 percent with individual temperature controls.[4]

The benefits of retrofitting buildings include boosts to environmental, social, and financial performance. Research shows that green buildings have the benefits of 8–9 percent reduction in operating costs, a 7.5 percent increase in the building value, and an increase of 3.5 percent in the occupancy ratio.

Whole-Project Financial Metrics

Some key criteria to help assess your project and return on investment include breadth, cost, knowledge, and collaboration. *Breadth*: how many people will your project impact? The greater the number of people, the greater the return on investment (ROI), because the single largest line-item expense for an organization is typically labor. This will help show the value of your high-performance, whole-building solution in the human terms of improved occupant health and productivity. The ability to build a dynamic business case for your project will get people's attention. The *costlier* the existing building is to operate, the greater the opportunity to benefit from the renovation project.

Buildings that are expensive to operate are perfect places to invest in technology and smart building infrastructure. This is especially true when improving building performance is done in accordance with the Natural Order of Sustainability: passive options first, active options second, renewables last. For those readers involved in managing portfolios of buildings (i.e., a college or corporate campus), the *knowledge* of understanding how to retrofit existing buildings along with the application of this knowledge across multiple buildings realizes economies of scale and can increase ROI. Finally, the greater the *collaboration* between owners, facilities, finance and IT groups, contractors, and occupants, the greater the return on your investment. This collaboration is at the center of an integrated project delivery that "integrates people, systems, business structures, and practices into a process that collaboratively harnesses the talents and insights of all participants to reduce waste and optimize efficiency through all phases of design, fabrication, and construction."[5]

With these contextual factors in mind, you will need to gather and validate data to support your investment decision. ROI is the most commonly used measure of average benefit over the time period divided by initial costs, so be sure to include monthly or annual cost savings and equipment-replacement life-cycle analysis of the proposed project. The payback period is how long it takes for benefits returned to equal the first cost of the project. This is a measure of risk, with the rule of thumb that sooner is always better than later. Net present value (NPV) is the value of ongoing benefits discounted back to the present year. This tells you if the project should or should not be undertaken. That is, if the project NPV is less than zero, you would be better off putting the money in the bank. It should not be compared to other projects unless the time horizon and amount of the investments are the same. Total cost of ownership (TCO) provides a good metric for understanding both your current costs of operating the building and the proposed future costs of ownership when the project is complete. Understanding these total costs will also inform cash flow in your NPV calculations. Cost savings and cost avoidance provide an important context for any project and should be looked at closely.

Consider the differences in cost savings and cost avoidance as part of an integrated approach to any project. Costs avoided are in the future, while cost savings are in the present. How these potential future savings are valued (and whether or not they include environmental and social performance) should be considered and can make a big difference in decisions to go ahead with a project. These costs constitute an integrated performance dynamic. At the lowest level, the base, we have cost savings. With cost savings, we address existing problems with low cost building materials and solutions built to code-based compliance. For example, we have an inefficient HVAC unit with diminishing capacity. We can apply a cost saving solution by purchasing a comparably sized unit. In doing so, the level of energy consumed and waste (e.g., GHG emissions) generated by the process could also be lowered by the new unit. It is important not to assume that the new unit will perform better, but to evaluate the impact of these changes by comparing the performance of the new unit with the level observed for the old unit. This performance data is necessary for validating capital expenditures and building a platform for data analytics.

Cost avoidance and tunneling through cost barriers is a better long-term and whole-system approach that yields bigger benefits than cost savings. Here you can go beyond traditional thinking in the left side of figure 4-1, where the more resources you save, the more you have to pay for each increment in savings. This is the concept of diminishing returns, and in this context, you should stop additional investments as you have made it to some limit of cost-effectiveness (the dashed line).

Whole-systems approaches to existing buildings move us to the right in this figure, where saving more energy comes from "tunneling through the cost barrier."[6] Taking this approach causes costs to come down and return on investment go up. When integrating engineering, design, and building modeling, bigger savings and bigger projects cost less up front than going for incremental cost-savings projects and their associated total costs (not first cost, which is all most people look at). There are two main ways to achieve this "more-for-less" result. First, integrate the design of all goals for the building so that investments in each performance metric result in multiple benefits (i.e., savings in both equipment and energy costs, as well as improvements in indoor environmental quality). The next way to approach this is to combine your efforts with improvement efforts for aging equipment, or façades that will need work in the near term. When taking action, optimize the whole system (the building), measure all benefits, and take the right steps at the right time, in the proper sequence, as laid out in our sequence of chapters. It's true that approaching the renovation of an existing building incrementally only requires an incremental investment; however, it does not allow an owner to access the significant cost savings or occupant benefits available by considering holistic investment analyses.

If you want to take a deeper dive into integrating the environmental and social value of your project, you can include the social cost of carbon (SCC) by placing a monetary value on every ton of CO_2e emissions avoided by the project. A typical number used is

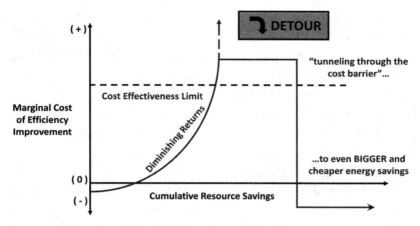

Figure 4-1. Tunneling Through the Cost Barrier (© Rocky Mountain Institute).

$40 per ton of CO_2.[7] When you can show that your project reno-
vation will save energy and avoid greenhouse gas emissions, you
can place a value on this using the SCC. Walmart, Disney, and Microsoft do this with their projects.[8] When applying an integrated
approach to developing a business case, we know from information
in chapter 2 that cost avoidance (future-looking high performance)
outperforms cost savings (present and short term).

Reducing whole-building energy consumption within an existing building is a logical place to start, and it's very easy to justify costs versus benefits. The dynamic environments of existing
buildings mean these energy projects typically involve lowering
water consumption, upgrading lighting, and improving indoor
air quality. When taking an integrated approach to understanding costs, you can reduce both first and total costs. An example project we have been working on involves Rockwell Hall, a
165,000-square-foot, ten-story building that is over fifty years old
(see the case study in chapter 8 for more details). A complete renovation of the entire building is not affordable due to capital scar-
city, so renovations will be completed on a floor-by-floor basis.
This staged approach does not prevent a holistic existing-build-

ing vision if renovations are coordinated under a master plan. A whole-building solution takes into account how each floor of the building fits with the vision of going beyond merely "renovating to code-based goals." Renovation plans can be made, materials sourced, and construction performed in a way that aligns each floor with a whole-building solution and performance assessment guided by third-party sustainability standards such as Energy Star, LEED, Living Building Challenge, Passive House, RESET Air, Fitwel, and WELL Building, among others (see box 1-2, "The Natural Order of Sustainability Approach," in chapter 1). Instead of a piecemeal approach to each floor, with the solutions focused on code-based goals, a holistic existing-building vision results in stepwise improvements of building performance and returns that take into account collaborative efforts to understand first costs and future benefits over longer periods of time. For example, if you know your optimum envelope solutions based on your building-wide goals, then you know your target R-value for walls and window performance. So you don't have to deconstruct an R-8 exterior wall and replace it with an R-8. At the same time, you don't have to remove a single-pane window and replace it with a triple-pane window because "it will perform better than the original." A master plan will identify the optimum envelope performance and prevent unnecessary overinvestment.

Construction Financial Metrics

Creating a compelling business case to renovate an existing building requires stakeholders to look beyond a system, a symptom, or a very squeaky wheel (human or otherwise). You have to be willing to consider the quality of the building operations and the impact the building is having on occupants. This chapter has provided a variety of ways to consider returns on your whole-building investment. It's essential, however, that we provide you with guidance on how to validate the inputs for construction costs for your calculations so that decision makers develop the kind of cost-confidence that comes from full transparency of the analysis. When we are advocating for owners, we want to provide them with all of the data they need to prioritize investments in existing buildings. We want enough data that, at any point in time, an owner can see the impact an individual decision or groups of decisions will have on building performance metrics, first (construction) costs and long-term operating costs. The table below shows you how we think about cost-confidence for construction costs. As a building owner, you should require a high degree of transparent cost-confidence from your project team as well as confidence regarding the lengths to which we will go to ensure project team members are providing full transparency into project costs. The key actions to validate financial inputs are listed below:

- ✔ Owner's Control Estimate
- ✔ Document Review and Analysis
- ✔ Investigation of Existing Conditions
- ✔ Subcontractor/Specialist Qualification and Review
- ✔ Price Estimates and Project Comparisons
- ✔ Budget and Order-of-Magnitude Pricing
- ✔ Cost-Management during Construction

OWNER'S CONTROL ESTIMATE

An accurate, comprehensive, and reliable cost estimate creates a solid foundation for any successful construction project. Every owner should have his or her own control estimate or budget, maintained confidentially throughout the project, which will serve as a guide for negotiations with all stakeholders during the project. A control estimate provides a basis for unit-cost expectations and labor production so that planners can know how much work is required for each task. A well-defined budget process and the experience of key personnel will increase owner and project team confidence in the control estimate and/or budget for any size and scope of project. The cost-estimating process should include construction-document review and analysis, investigation of existing conditions, review of subcontractors and specialists, conceptual and detailed price estimates, and process controls for cost containment during construction.

DOCUMENT REVIEW AND ANALYSIS

Cost-estimating includes first itemizing and quantifying a detailed scope of work through quantity takeoffs and other analytical means. Identify and organize a scope of work based on the construction standard index or work-breakdown structure codes and generate recap spreadsheets grouping related items. Also consider a performance-constructability analysis on all construction documents as part of preparing an overall estimate. Constructability analysis considers how efficiently, effectively, and safely a process and phase of construction can be executed. Constructability analysis should also include drawing detail reviews for thermal bridging, thermal and air-barrier continuity, and mechanical and electrical systems. Try to examine all aspects of the construction in an effort to reduce cost and schedule duration by looking at ease of construction and alternate materials and methods. Constructability reviews typically include project phasing and duration, selection and coordination of systems, coordination of drawings, construction details, site logistics, environmental concerns, submittal requirements and long lead materials and equipment, the safety and security requirements of all phases of work, design conflicts, temporary utilities, and construction quality management. Time spent performing constructability reviews will go a long way toward assuring both a well-coordinated schedule and a minimized impact on ongoing operations.

INVESTIGATION OF EXISTING CONDITIONS

A qualified member of your team should perform a site visit prior to completing the cost estimate. A thorough site investigation is necessary to clarify any general scope items. On many projects, there is a scheduled pre-bid meeting and/or site walk through with the owner and design team. These meetings help establish guidelines for the construction and allow all parties to elaborate on the project requirements outlined in the design. We put a specific emphasis on maximizing this opportunity to review the project site and ask questions regarding the construction documents and any project specific requirements.

Subcontractor/Specialist Qualification and Review

In an effort to accurately estimate and price a project, provide construction documents to various potential subcontractors and suppliers in order to solicit additional pricing for specialized scopes of work. When possible, utilize a variety of methods for document distribution through cloud-based solutions. This approach may vary depending on the security sensitivity of the documents or the capabilities of the other parties. Review each subcontractor and supplier proposal for scope, content, and pricing accuracy.

Develop your own customized procedure for qualifying subcontractors and verifying the scope of work prior to awarding contracts. During the estimating process, generate a scope-of-work review sheet for each trade to enable comparison between each subcontractor's proposal. Finally, you will want to coordinate an interview/scope meeting with potential subcontractors in which the project manager, estimator, and probable superintendent participate.

Price Estimates and Project Comparisons

Each scope of work should be itemized and priced on a unit-cost basis. If you do not have expertise in doing this, look for help from those who have established production-rate information for all trades. Trades can be priced using both analysis of historical costs and accurate, up-to-date material and labor pricing. Upon completion of a recap spreadsheet for each scope of work, create a final spreadsheet which groups, organizes, and totals all recap sheets.

You will need estimates for indirect costs of construction (commonly referred to as general conditions/general requirements) on a separate general conditions sheet. The general conditions sheet values all indirect costs on a project-by-project basis, and these costs vary according to project duration and complexity. Typical general conditions sheet items include mobilization, job-site staffing, temporary facilities, temporary construction, utilities, construction equipment, small tools, consumables, testing, and other categories. All costs can then be transferred to the final spreadsheet. The final sheet also itemizes the costs relating to the plant, labor rates with burden, fees and markups, contingencies, and project bonds.

BUDGET AND ORDER-OF-MAGNITUDE PRICING

On occasion, a thorough and complete estimate is either not necessary or not possible. We are often faced with the need to predict the future cost of a conceptual project with little definition in terms of design or building performance. When this happens, look for experts with significant experience preparing a wide variety of budgets and rough-order-magnitude estimates. These estimates are useful for project planning and design/build delivery methods, or for simply determining whether a complete estimate will be useful.

Buildings are complex entities with millions of parts. Therefore, predicting the cost of an incomplete or yet-to-be-designed structure or scope of work is as much art as science. Balancing this by benchmarking costs early and often is a useful strategy for managing expectations and outcomes. This benchmarking process is a careful and collaborative one; therefore, we do not simply turn drawings over to a contractor/estimator. Instead, try to work closely with contractors, engineers, suppliers, and clients to determine cost ranges for a project that are appropriate to each scope of work. If necessary, conduct budget workshops with various members of the project team in an interactive, open-book format in order to engage the entire team in the process of budgeting. In our decades of experience with owners of all types, sizes, and financial capacities, we have seen firsthand the benefits of this collaborative approach. Seek out those with the necessary knowledge and experience to generate an open and transparent budget tailored to the specific needs of your project.

COST-MANAGEMENT DURING CONSTRUCTION

We have managed project budgets using a combination of Sage Timberline Office, Planswift, RIB MC2, Oracle Primavera Contract Management, and Microsoft Excel. We consider this to be the final part of the estimating process, and it is an important component in successfully managing a project. Create a detailed project budget based on the cost estimates that will be used to manage project goals and priorities through the construction process. All budgets for individual aspects of the project should be continuously re-evaluated to ensure that the project is constructed within the established estimate and to the satisfaction of end users. As necessary, adjust the budgets for individual aspects of the project to accommodate different design options and end users' requirements.

Performance beyond First Costs

Strategic alignment involves value maximization. The process for cost estimation as outlined above puts you on the path to ensuring a strategic approach to existing-building project estimation. We can achieve cost savings and cost avoidance without even considering issues such as identifying key stakeholders or fitting into the business model of the enterprise. To increase the value associated with the existing building, you can also consider the environmental, social, and governance performance of the building project. This means that we look holistically at how buildings help create value. When dealing with value creation and the enterprise's business model, we have changed the focus of integrated high-performance buildings from operational or tactical to strategic. This level of thinking is the reason that a higher level of integration (value maximization) is one in which we can simultaneously account for performance beyond first costs in decision-making.

When viewed from this perspective, environmental and social performance can be integrated into and simultaneously contribute to business sustainability. Innovative projects can start with a solid business case, yet also need a compelling story about the integrated value they can deliver. This is a vision that focuses both on value creation and a reduction of total environmental and social impacts. It represents the benefits to building operators, occupants, and even the enterprise's customers. It represents a situation where whole-building solutions are critical for the enterprise and the environment, and are viewed as essential to developing and maintaining a competitive advantage.

When looking for opportunities for strategic alignment, review the evaluation criteria for your project to identify opportunities for value creation, along with hidden costs. Look for solutions to problems instead of merely addressing their symptoms. For example, look at what you can do with whole-building performance, energy conservation measures, and indoor air-quality metrics to gather building intelligence. Plan to measure pre- and post-project performance. Share the results with building stakeholders and find ways to translate success into messages delivered to occupants and customers, communicating how your organization is different from those that choose not to see the strategic alignment that existing-building projects have to offer.

Those who are willing to take on the challenges of whole-building transformation will be able to understand the total costs and full returns on their capital expenditures. We just need to start asking the right questions. What is our current site EUI? What are our current IAQ trends? Are these results good or bad? How can we improve these whole-building metrics? This knowledge will place you at the cutting edge of the dynamic capabilities of integrated, whole-building solutions. For these types of efforts, innovative project teams should be celebrated as the change agents for which the world has been waiting.

We see this as a call to action for the next generation of building owners and building occupants to truly understand what it means to measure and manage building performance with the strategic goals of maximizing the value that existing buildings can provide. As we issue this challenge, we know it comes with pitfalls and barriers that at times can seem insurmountable. By preparing to support your existing-building vision with standard business language, building science, and simulated data, you and your team will show the need and financial feasibility of your project, leading all those watching from the bleachers to ask, "Can we afford *not* to do this?"

Funding Strategies

Opportunities exist today for nontraditional funding for existing buildings that are attempting to reach high performance. For example, your state level Department of Community and Economic Development (or some equivalent) may support high-performance building programs, providing low-interest loans and grants for the design and construction of major retrofits of buildings. Small businesses and individuals are eligible for these loans and grants. In the case-study chapter, we will look at the Three Rivers Luxury Residential Tower project. A condominium association like Three Rivers Luxury Residential Towers would benefit from pursuing DCED funding. While not technically a small business or an individual, Three Rivers Towers represents exactly the sort of solution that the DCED mission is trying to accomplish in supporting the design and construction of major retrofits to turn old buildings into high-performance examples of what can be accomplished with existing buildings.

There are other organizations like sustainability energy funds (SEF), Better-Building Energy Efficiency (BEE), and Metered Energy Efficiency Transaction Structure (MEETS) that specialize in bringing financing to projects that use Passive House principles to reach low energy. Many owners of existing buildings choose to avail themselves of the services of energy savings companies (ESCOs). ECSOs serve very useful purposes but we caution against bringing them into a project before you have full transparency and control over your performance data and expectations. Without proper guidance, ESCOs will determine their program based on the most financially attractive energy-conservation measures for them, not the owner. Because they provide the capital for the project investments, their investments are paid back handsomely. In our opinion, the best use of an ESCO is to invite them in to develop a funding proposal once the owner has determined the priority of investments that are in the owner's best interest. For any owner looking to access 100 percent of the benefits of the investments in renovation, we suggest finding sources of funding other than an ESCO. Their business model requires them to harvest all the low-hanging financial fruit first.

Next, cities or municipalities in which your building is located are not without capital or resources, but it takes political influence to get to the proper decision makers. For example, organizations like redevelopment authorities should want to be involved in existing building projects. Their mission is to support anything that drives "a thriving and sustainable city." They typically have state and federal funds, and they invest in what they think will make the biggest impacts on the city. In our opinion, there is nothing more accessible in terms of "difference" than restoring existing buildings.

Lastly, adding a professional grant writer to your project team provides you with experience, political heft, and a greater level of sophistication to help find new sources of funding. Your team should tell the opportunity story of your building and talk directly with funders about the possibilities of and vision for its high-performance renovation. Many cities have very experienced grant writers specializing in the built environment. A word of caution, though: make sure you check references from their most recent work.

When starting a conversation about a project, cost savings is always a good foundation on which to build your ideas and performance-measurement opportunities. Then look for cost avoidance and strategic alignment as you develop a complete plan with defendable financial objectives.

Project success is typically determined by the completeness of the planned deliverables (a scope of work), delivery according to a schedule (time), and meeting financial objectives (the business case). The incredible opportunity within existing buildings means that, with a plan and knowledge, the potential amount of financial benefit is staggering. We now know that a do-nothing, business-as-usual approach is frustrating, irrational, and wasteful. Why keep wasting resources on poor-performing buildings? Instead, transform existing buildings to high-performance buildings with low energy consumption and superior indoor environmental quality without spending a premium in construction costs. Zero-energy goals are within reach. Transforming existing buildings to high performance with the potential to reach zero-energy requires modern building science and technology solutions. The next few chapters will explain the challenges presented by the world's stock of existing buildings, why the envelope is one of your most important considerations, and how technology makes it easier for stakeholders to "just say yes."

As Benjamin Franklin opined, "An investment in knowledge pays the best interest." When it comes to investing, nothing will pay off more than educating yourself. Do the necessary research, get the building baseline data, and study and analyze it before making investment decisions.

Best Practices

Finally, we suggest a few best practices that we have found to increase the likelihood of meeting building performance goals in operations:

1. *Use Open-book cost estimating.* Any project with aspirational building performance goals should be conducted using an open-book approach.
2. *Empower your performance advocate.* Make your performance advocate the champion of the whole-building performance model. This ensures that the model and the data associated with the model are always accessible and available to all team members.
3. *Maintain your own project-control budget.* We believe that owners have a great deal more influence on projects when they have an independent cost-control estimate executed at all major stages of construction. An independent estimate combined with an open-book project approach provides owners with the greatest control over project costs and the potential trade-offs that are created on every project. It's important to note that an owner is not required to share the independent estimate with anyone on the project team. Typically, the performance advocate oversees the estimating process to ensure that it is unbiased.

Conclusion

Knowing that your building has enough potential for cost savings and waste reduction provides a basis for believing that you can have a different vision of the future. Armed with a plan, baseline data for your current performance, and clear-cut goals, you are now prepared to develop a holistic investment approach to transforming your building. The development of a business case for your project, alignment with strategic objectives, and the involvement of a team are critical to ensuring the success of any existing-building project. You should approach the business-case development as a way to leverage your organization's strategic priorities. In doing so, the full impact of integrated systems that include environmental and social performance will bring your efforts and your existing buildings into a new modern era of performance.

Project Development Homework:

Consider a traditional first-cost approach to any planned renovation to analyze against a whole-building, deep-energy retrofit approach. Then, take a step back and look for all the benefits your renovation can provide, and include this valuation in your ROI, NPV, and TCO. Can you stretch your goals even further to a zero-energy solution, or a lowest-energy approach to the integrated design?

- ✔ How much money can you save if you transition from a first-cost, cost-savings approach to the project to a cost-avoidance and value-maximization approach?
- ✔ Identify and rank the pitfalls and barriers you may encounter when developing and implementing your whole-building design.
- ✔ Run a design-charrette / scenario-planning exercise with your team. What are the worst-case and best-case scenarios for your existing-building project? Acknowledge the risks, and then capture the long-term integrated financial, environmental, and social benefits of the project.
- ✔ If you know how much your energy consumption can be reduced, find the tonnage equivalent of CO_2e and apply social-cost-of-carbon (SCC) calculations to show the impacts avoided, and thus the value created, by your project, which takes you beyond a single bottom-line, first-costs approach to decision-making.

Chapter 5:

The Building Envelope Holds the Key

*Someone is sitting in the shade today because
someone planted a tree a long time ago.*
—Warren Buffet

W E ENVISION A FUTURE IN WHICH society will routinely
consider restoring existing buildings to the performance standards of new ones—whether an existing building is abandoned
and forgotten or outdated and underperforming. We hope that,
over time, it becomes socially and financially unacceptable—
even unfashionable—to build a new building when there is an
existing building perfectly suited for restoration in the same
neighborhood. The pace at which we achieve this aspiration
depends on the urgency we show to expand design imperatives
beyond mechanical system components to include envelope
system components. The quality and performance characteristics of a building are fundamentally defined by the quality and
performance characteristics of the building envelope, which
is often overlooked by the design teams on existing-building
projects.

Performance in Operations

As we like to say, "It's no longer enough to design ultra-low-energy, ultra-high-indoor-air-quality buildings. Owners today expect performance in operations."

Unfortunately, the horizontal nature of traditional building projects doesn't lend itself well to hard-wiring the conceptual design phase to operations. This chapter takes the first step toward building the confidence of building owners and developers that the expected results will be delivered in operations. This is essential if we are to convince key decision makers to invest in existing buildings.

To do that, we focus first on the importance of designing a high-performance building-envelope assembly. Project teams often spend most, if not all, of their time considering and planning how to change or replace HVAC and mechanical systems. Very little, if any, time is spent on the building envelope. The first key to controlling building performance is to control the building envelope.

It is almost impossible to make existing buildings perform like new without considering the building envelope. Watch out for naysayers proclaiming, "The building envelope never pays!" Many traditionalist engineers say such things, but we disagree. Investing in continuous insulation and good air and moisture barriers are much more beneficial and cost-effective than investing in more heating and ventilation to simply pump it all through a leaky building—right? Insulation is always cheaper than mechanical systems and is far more durable and lasting.

Whether we are talking about restoring a forgotten building or updating an outdated one, the building science is the same. Reaching a building's theoretical optimum performance will never be achieved without optimizing the performance of the building envelope.

The most sophisticated building science standard focusing on the importance of the building envelope is the Passive House standard, first introduced in chapter 1. Passive House relies on increased insulation, airtightness, energy-efficient lighting and equipment,

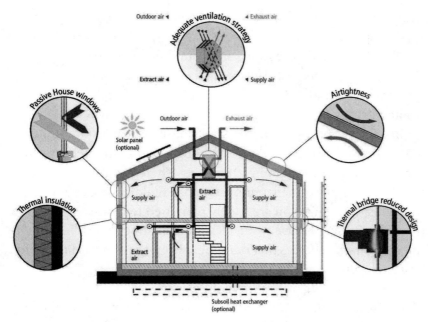

Figure 5-1. The Passive House Institute Envelope (© Passive House Institute).

and improved ventilation to hit its aggressive performance targets for whole-building performance (see fig. 5-1). To reach the type of building performance expected from Architecture 2030 Challenge goals or Zero-Energy, the passive system of the building envelope must be properly designed for high performance.

Passive House is a set of design fundamentals and principles that creates thermally efficient and airtight structures. The results are lower energy consumption and lower costs for both heating and cooling. The building itself is used to regulate indoor temperatures. While not originally designed for indoor environmental performance, Passive House buildings have, ironically, led to enhanced health, comfort, productivity for occupants, employees, and residents.

In fact, the world's first Passive House–designed hospital is under construction in Frankfurt, Germany. The six-story, full-service, twenty-four-hour hospital will dramatically cut energy consumption and will increase thermal comfort for staff and patients. The project is expected to reduce energy by 40–60 percent From implementing a high-performance building envelope to sourcing high-efficiency medical appliances, Passive House design strategies provide the foundation to drive low-energy, and high indoor environmental quality even for the highest-load buildings.

In sharp contrast to the Passive House principles are building codes that do not ensure that buildings are constructed or renovated with high-performing envelopes. There is a running joke in the building-performance industry that building codes represent the worst buildings you are legally permitted to construct. The gap between Passive House principles and building codes lies fundamentally in the choices that building owners make regarding building enclosures.

Typically, strategies to retrofit an existing building start off with deferred maintenance or with systems or components at the end of their life cycle. The focus is often then on replacing or fixing systems or components in kind. The envelope is normally not part of the discussion during routine building-maintenance planning. Roofs, doors, and windows are typically replaced in kind, which has little impact on overall building performance. Unfortunately, the structure of the construction industry has unwittingly conspired to keep high-performing renovation strategies for existing buildings out of the hands of interested and enlightened owners.

Architects are the key to high-performance envelopes. But, because most planning regarding existing buildings tends to start with mechanical and electrical engineers, architects aren't always the first team members consulted, unlike with a new building. When building owners spend months of work and lots of money with mechanical and electrical engineers or performance contractors (i.e., Energy Service Companies [ESCO]) determining what should be done, consulting an architect simply does not seem to be economically justifiable. In the worst situations, an architect is never consulted at all.

But today we know much more about the relationship between building performance and the envelope, knowledge that we cannot ignore anymore. Architect Laura Nettleton is one of the founders of the Passive House Western Pennsylvania organization and a renowned leader in achieving breakthrough building performance without spending a premium in construction costs. In her words:

> It comes down to commitment to goals. If you are really committed to reaching low energy use and/or high indoor air quality, and you make those priorities for your project from early conceptual design, then pursuit of Passive House will cost very little, if any, extra money. However, if you try to force Passive House goals on a project that is already designed and detailed, pursuing Passive House design or certification could cost as much as 10 percent more than traditional construction. Even if a project is well into design, the team may still find that the cost premium of Passive House is justified based on the energy savings alone. The idea, parroted back to me from other architects or engineers, that pursuing Passive House can lead to 20 or 30 or 40 percent increase in costs, is simply nonsense.

How do you want your building to perform in operations? If you want to reach high performance in operations, it is important to set goals very early, in the form of specific performance metrics. We agree with Laura Nettleton that the key to managing costs on a project is to get very specific about building performance goals early in the design process. The most aspirational building-performance goals require teams to first optimize the envelope which begins to give the owner the necessary confidence that the building will deliver high performance in operations.

If you believe that a high-performing envelope is a critical success factor for your project, you should consider the following points regarding the on-boarding of an architect.

Selecting an Architect

To determine the full potential of an investment in the envelope for your project, find an experienced Passive House architect. They will be able to design the highest performing envelope for your building, given its constraints. They are also responsible for bringing other design professionals to the project. There are at least 10,000 different decisions to make on every project. Along with those decisions comes a hierarchy of decision-making. What is most impactful and what is least impactful? Even in the worst case, Passive House standards may still be worth pursuing, because the return on energy savings and improved air quality typically provides an ROI of two to four years, depending on how you calculate those returns. In every situation, you should run the numbers, as discussed in chapter 4, before ruling out investing in a high-performing envelope.

If you do not select an architect who has Passive House experience, then owners will find themselves fighting the design team on high-performance strategies, which is what causes change orders, schedule delays, and inflated cost estimates from built-in "fear" premiums. People have all kinds of misconceptions about what defines high-performance buildings, such as the belief that you can't have lots of windows, or you can't open the windows, or your building will not breathe, or your green goals will cost too much, or you must sacrifice aesthetics. All those misconceptions are hogwash. You need to prioritize high performance right from the beginning.

This idea is illustrated beautifully in examining the Passive House project costs of some affordable housing projects that were awarded low-income housing tax credits by the Pennsylvania Housing Finance Agency. In 2015, PHFA began to award projects for achieving Passive House standards. Passive House moved up on the hierarchy of the owner's decision-making matrix and the number of projects seeking Passive House certification increased. It is safe to assume that not many project teams had prior experience with Passive House. In 2015, Passive House projects cost about 5.6 percent more than traditional construction. In 2016, Passive House projects posted a modest 1.6 percent premium over traditional construction. However, in 2018, as project teams deepened their experience with Passive House, costs for Passive House projects averaged 3.3 percent *less* than traditional construction.[1] With experienced teams and owner/developer commitment to Passive House standards, projects should not result in any premium in costs.

Drivers of Innovation

Two enablers in building design created the Passive House movement and the rise of high-performance envelopes. These enablers were the growing understanding of the importance of air infiltration and exfiltration and the advancement of building-performance modeling technology.

Historically, architects considered continuous insulation as the primary factor for preventing heat loss, and the R-value was the litmus test of performance. As awareness grew, however, it became apparent that air infiltration and exfiltration are every bit as important. If you insulate the roof on a single-family house with R-50 insulation but you haven't properly sealed around the plumbing vents, chimney, and other protrusions, your R-50 roof might not perform to an R-50 level. Forensic evidence tells us that those areas that are not sealed properly are likely to be the first places to show signs of building failure. Moisture in the air migrates through those weak spots, and mold and moisture gravitate to them. In fact, failures in the building envelope rapidly became the focus of high-performance and envelope-durability studies. Airtightness was quickly understood to have an equal level of importance with continuous insulation and is now considered a key to building performance.

Innovative technology gave architects, engineers, and construction professionals a chance to understand their buildings and their building's performance in ways not previously available.[2] While we have understood for a long time that heat migrates through the building's skin, taking moisture with it, we are now able to calculate performance impacts. In the 1980s, during the early passive-solar movement, architects used handheld calculators, protractors, and sun-angle calculators to determine sun-path diagrams, dew point, length of overhangs, and heat loss through the skin. But there were far too many discrete components of the building to do calculations practically for each one.

When computers became fully integrated into architectural practice, computerized energy modeling was able to generally evaluate the impacts of a building's orientation, wall assemblies and windows. Some of these models (IES VE, EnergyPlus, eQUEST, TRACE, Hourly Analysis Program [HAP], and others) are more precise than others, but all are useful for energy performance and/ or sizing mechanical equipment. In the early days of LEED, these results were compared to a base model (ASHRAE 90.1) of a building designed to code, and then they were compared against an upgraded design in order to predict performance savings. This resulted in a generation of buildings that aspired to reach low energy consumption but did not always realize their goals. Building owners became disillusioned with the prescriptive rating systems, which promised performance but didn't always deliver, leading to skepticism about the possibility of truly reaching high-performance goals.

As computer technology advanced, new energy-modeling software allowed designers to understand and calculate each component part, which for the first time enabled them to *accurately predict* the performance of multifaceted envelope systems. Technological advances in meters and sensors allow building owners to *continuously monitor* the performance of their buildings.

A new understanding of envelope design, coupled with new methods of quality assurance / quality control in envelope construction, makes more precise prediction of building performance possible.

A New Era of Existing Buildings

Now that the tools are available and the details are all in the control of the design team, there is little or no reason for new construction not to pursue high-performance targets in both energy consumption and indoor environmental quality. It is a relatively simple process for any experienced design professional, and there is usually little to no premium in construction costs. In fact, pursuing high-performance targets is the only responsible course of action.

However, an existing building should always be evaluated by the design team in order to ensure that it is a good candidate for a high-performance retrofit. Some buildings have important social and historical value that would be unacceptably altered by extensive internal or external renovation. Some building owners and communities want to maintain the original look and feel of an existing building. In such cases, insulating from the inside is preferable. In fact, it may be required by zoning or other local ordinances. In other cases, existing building construction, thermal bridging or occupant conditions limit insulation choice to an outside approach, irrespective of owner or community preference.

If the building has a tired and dated image, that image may be depressing the market potential of the building. Public housing projects from the 1950s and 1960s, for example, represent a time of failed social policy, even though they were generally made from quality materials. In this case, insulating from the outside is the optimal choice because the building's value and image can be redefined. The new cladding and insulation are like putting a sweater over the building. With its new "outfit," the building has enhanced market appeal, lower utility costs, and superior air quality. Insulating from the outside is generally an easier process than interior insulation, as the team can control all the existing thermal bridges and other complexities from the original building. Projects that choose to insulate from the outside typically have an easier time contending with the details in design and construction.

There are even cases where high-performance retrofits are done from the outside while residents and building occupants continue to use the inside spaces. A new skin is placed over the old one, the windows are replaced, and the project has very little negative impact on the residents' lives. Energiesprong is a Dutch "whole house" refurbishment system that reclads and reroofs buildings with residents in place. The program began in the neighborhoods of the Netherlands where housing is largely standardized, but the approach holds great promise for other exterior retrofits. Using a laser scanner HDR camera, the existing conditions are mapped in their entirety, and the information is translated into a digital image. That image captures the exact measurements of the building with all its imperfections. The digital file is then translated into a shop drawing for the modular panel fabricator at the factory. Each modular panel is numbered so that when it arrives at the site, it can be installed in the proper location. The panels are craned into place and sealed, making all the construction relatively quick. Roof panels are handled in a similar way, and so the house gets a whole new skin along with its energy savings.

Insulating existing buildings from the inside or outside can be tricky, so hiring an envelope consultant is essential for retrofit projects of this kind. The envelope consultant can evaluate the condition of the building façade to understand whether the existing cladding can withstand the new condition when not being heated from the inside. Some masonry walls are not able to withstand the cooler temperatures and may lead to masonry unit and envelope failure. Understanding the nature of the material is key and may be beyond the expertise of the architect. If the design team is insulating an existing building from the inside, it is worth getting a consultant involved to understand how it will perform under these new conditions. This is a must for anyone trying to insulate an existing building.

Frequently, minor selective exploratory demolition may be required in order to understand the existing building conditions and fully assess a building's potential for envelope restoration. The team may not have addressed envelope issues in past renovations and may have unknowingly neglected vapor migration through the building skin, unwittingly affecting the building's durability. There may be mold, dampness, or deterioration of the building components. Ideally, a Passive House "EnerPHit" strategy is undertaken for existing buildings whose windows and systems are at the end of their useful lives. Bundling numerous life-cycle and deferred maintenance projects into a larger project to overhaul the entire building is typically the best solution. However, building owners are not always able to replace every component at once. Sometimes buildings have to be upgraded in a piecemeal fashion. In this instance, a step-by-step EnerPHit approach can be taken, as warranted and as money becomes available, with the ultimate goal of reaching the Passive House standard. For example, if windows require replacement, then the windows may be added in a manner that allows the air barrier to be continuously connected into the window jambs at a later date. The parts are installed separately so that they do not compromise the goal of airtightness, leaving it open for the air barrier to be continuously installed as time and money permit. The planning discussed earlier is based on our analysis and learning from Rockwell Hall.

It is important to consider the needs of building owners when embarking on a deep energy retrofit. Generally, whole-building retrofits are only pursued every 50 to 100 years. Building owners usually have multiple goals for any major retrofit. They want to update their buildings, but they also want to improve their buildings' function. Perhaps there is additional programming or improved lighting; or maybe they wish to install amenities such as elevators, automatic entrance doors, air-conditioning, computer rooms, dishwashers, garbage disposals, laundry facilities, or TV and data ports. The building code itself may require fresh-air ventilation systems, fire-sprinkler systems, upgraded electrical service entrances, and life-safety systems that all require more energy. Perhaps the owner wants to implement changes that have the potential to lower energy consumption, like demand-control ventilation strategies that account for occupant variability. But, typically, the desires for the new building increase the energy consumption for the new project. Including these elements in the discussion of goals is important, as the increase in energy consumption can be significant.

Building-Performance Modeling

Now that architects have factored in all the options, they will provide a massing and orientation for the building that will include a high-performing envelope design. With that data, you are ready to move to the next critical piece of the high-performance puzzle, the whole-building performance model.

To construct a high-performing envelope, the project design team needs to understand the site conditions. The eight climate zones in the continental United States are all featured in high-performance building modeling, which must take into account humidity, wind, number of degree days, sun path, shade objects, solar availability, orientation, temperature, and specific site features. Other considerations are occupancy patterns, window placement in the envelope, shading devices, glazing characteristics, window-frame construction, continuous air barrier, weather barrier, thermal bridging, daylighting, lighting, HVAC equipment, domestic hot water, and other pieces of equipment. An experienced design team knows that each one of these factors needs to be considered in order to bring up a successful project.

An architect specializing in high-performance buildings and having Passive House experience may be able to model the project in-house. This has advantages as it relates to envelope design, but not many architects have the capabilities or experience to do this kind of modeling unassisted. If your design team does not have this capability, then it is essential to find an experienced Passive House practitioner to support this modeling work. Irrespective of who undertakes the modeling work, it is essential that the model and resulting reports be made fully available to the entire design team. Our philosophy is that the model is the property of the building owner and should be made available to all team members in order to enable design teams to see the implications of design decisions from a performance standpoint. For example, can we swap out some of that expensive under-slab insulation for more insulation on the roof in order to save money? Trade-offs are easily evaluated and integrated into the project to stay within the target budget.

The timing of the building-performance model in the design process is important to understand. If your team is starting the energy model at the end of schematic design, it may be too late to be of optimal value from a performance standpoint. A rough model should be started alongside the early conceptual design in order to understand design-decision impacts. Issues you might consider early in the process are envelope optimization, solar orientation, shading, size and orientation of windows, and building location on the site. As designers gain more experience with building-performance modeling, they begin to learn about the consequences of their decision-making and those lessons become incorporated into their growing body of knowledge and experience. The first project using Passive House building science can feel daunting, but after a project or two, the designer's ability to predict outcomes improves with experience.

Integrated Design Process

An integrated approach to the design process and modeling is best. There is some confusion in the design and engineering community about what an integrated design process really is. Integrated design invites the building occupants and the design and construction teams to work together to make a project fit the budget and serve the owner's needs. Simply having everyone at the table is not enough. Integrated design means listening and incorporating team members' input into the design from the outset. For example, if there is a structural limitation that will significantly add to construction costs, then that knowledge must influence the design. Similarly, if the utilities come in at one side of the building, locating the mechanical room on that same side of the site will help reduce costs. If the building occupants cannot use a space because it does not have certain amenities, then those items must be included in the design or the project won't be used to its greatest advantage. An integrated design process takes the knowledge from every design team member and from the stakeholders on the owner's side and uses all of this knowledge to shape the project decisions. Using the performance model ensures that the loudest voice in the room does not dominate. In fact, *every* voice in the room carries equal weight, with empirical evidence driving decisions. In a sense, there is one truth backed up with data.

Design of Building Components

In high-performance envelopes, the devil is in the details. Because airtightness is so essential, the detailing determines the success of the project. If an air barrier is continuous around the building (under the slab, up the walls, into window jambs, and over the roof), then the transitions of the air barrier are critical to prevent air infiltration and exfiltration. The architect's drawings should reflect each of those conditions and the transitions between sections of the air barrier.

WINDOWS

Windows in high-performance buildings are typically triple-glazed and have insulated and thermally broken frames. R-values for high-performance windows are in the R5–R7 range (historically, high-rated windows in the United States average R1). This ensures thermal comfort for building occupants and minimizes condensation between the windows and the adjacent walls. Other specific characteristics of the glazing, such as solar heat gain coefficient (SHGC), U-value of the glazing, window-frame construction, and visible transmittance, all play their part in the energy profile. Even window screens influence performance.

AIR-BARRIER MEMBRANES AND TRANSITIONS

The air barrier in a high-performance building is continuous, like a balloon. This continuity is essential for performance. The air barrier takes a variety of forms, depending on the type of wall assembly and its location in the building. Some membranes allow moisture to migrate through the skin, and others do not. Determining the type of air barrier and whether it is permeable or not are decisions based on the hygrothermal analysis of the wall assembly designed by the architect. Membranes come in sheets or in self-adhering products, and some are fluid-applied. Material type has implications for cost, installation time, and reliability. Some materials are more suitable for some locations than others. Let's assume you have sheet material underneath the foundation and you are trying to tie it into a fluid-applied wall system; you must have a design detail in the drawings for transition between the two materials and possibly specifications for an additional material. Transition details and material compatibility are every bit as important as the material itself. The architect must include all the materials and details in the drawings.

THERMAL BRIDGES

Thermal bridging occurs when a more conductive (i.e., poorly insulating) material allows an easy pathway for heat flow across a thermal barrier. A common form of this is found in wall studs and structural steel framing. Mitigation of thermal bridges is another important issue that your architect should include in his or her details. In traditional buildings, it is common to find the structure running from the interior of the building to the outside. The outdoor air condition travels through the structure, creating a vulnerable spot in the envelope that allows for condensation to form and allows the cold outside to transmit to the warm inside. Thermal bridges are not to be underestimated and can account for 30 percent or more of the heat loss from a structure. Designers who are thoughtful can isolate different parts of the structure and can minimize or eliminate thermal bridging. In many cases, thermal bridging in an existing building can be mitigated, if not eliminated altogether. The structural engineer is often involved with these decisions and can play a vital role in minimizing costs if brought into the discussion early.

Mechanical Systems Design

Just as it is critical to select an architect experienced in high-performance techniques, it is equally critical to require that kind of expertise from your mechanical and electrical engineers. The first goal of a high-performance building is to invest in a high-performance envelope. This will translate into reduced building loads for heating and cooling. Lower building loads permit the designers to use smaller and decoupled mechanical systems. The money saved on reduced mechanical systems will most likely offset or surpass the money spent on the envelope. Traditional design practices include large, elaborate mechanical systems that are oversized to compensate for the leaky envelope. Engineers historically managed poor envelopes by pressurizing the interior spaces as necessary to prevent air infiltration through a leaky envelope. More alarming, many facility managers justify leaky envelopes of their existing buildings as a benefit by providing "make-up" air for ventilation and air movement. Obviously, this make-up air is not filtered or treated before it enters the occupied spaces and breathing zones.

The building is ideally seen as a system of passive and active components working together. The predictability of the envelope performance makes the downsizing of equipment now possible. Ventilation air in high-performance buildings is usually decoupled and supplied in a separate system from the heating and cooling supply. Experienced engineers will understand how to minimize expensive duct runs and how to strategically place equipment for better performance. Traditional design practices add wide margins of error because they cannot accurately predict performance. In a high-performance building, the project team can accurately predict performance of the envelope so that the equipment can be right-sized for the space it serves, with a much smaller margin of error. This relies on the precise control of the envelope through enhanced commissioning.

Communication with the Contractor

Good communication between the design team and the contractor is an essential component of delivering a high-performance building and its envelope. The architect needs to clearly define the intent in the drawings. Typical drawings clearly delineate the air barrier and its location and transitions in the building section. This requires a description of the intent of these building-performance components. Some architects print their documents in color and make the air barrier red to highlight its importance. It is advisable to have a preconstruction orientation meeting with the architect, contractor, and subcontractors to discuss and plan the installation of the envelope assemblies. Many subcontractors will not have worked on a high-performance building previously, and they need to understand the project goals and the importance of the air barrier. They may be used to drilling through the envelope, but high-performance aspirations make that unacceptable. Issues will arise in the field that challenge the project team, so knowing when to call the architect is critical. Everyone needs to be on the same page, dedicated to the goal of airtightness and preservation of the air barrier. In existing buildings, thermal imaging and smoke sticks can help identify problem areas during air infiltration and exfiltration testing so that the quality of the project is maintained.

The 3-30-300 Rule

For most organizations, facility managers know instinctively that the costs associated with building occupants are the highest costs. The relationship between energy-consumption costs, rent expenses, and building-occupant costs are summed up in the 3-30-300 Rule, as follows:

$$\$3/sf = \text{Energy Consumption} \qquad \$30/sf = \text{Rent}$$
$$\$300/sf = \text{Building Occupant}$$

When we consider the impacts of indoor environmental quality on building occupants, any investment in the building envelope is financially inconsequential. Indoor environmental and air quality is broadly understood to have a meaningful impact on human health and productivity. Note the following results of empirical studies:

- On average, cognitive scores were 60 percent higher in green-building conditions and 101 percent higher in enhanced green-building conditions," as reported by the Harvard Center for Health and Global Environment in "The Impact of Green Buildings on Cognitive Functions."[3]
- "Exposure to residential dampness and mold contributed to 21 percent of 21.8 million cases of asthma each year," as reported by the Harvard School of Public Health in "Building Evidence for Health: The Nine Foundations of a Healthy Building.[4]
- Childhood asthma is a leading cause of student absenteeism and accounts for 13.8 million missed school days each year, according to the Centers for Disease Control and Prevention (2015). With increased absences, a student's test scores may begin to reflect less how much the student studied and more the student's health and ability to focus on learning. This is reported by the Harvard School of Public Health in "Schools for Health: Foundations for Student Success."[5]

Building Guidelines and Regulations

As our industry spent most of the past decades innovating with all forms of mechanical systems and renewables, we lost sight of the importance of the most basic functions of a building: keep the outdoor conditions outside and the indoor conditions inside. The envelope is the key to these basic functions. As we endeavor to make old buildings perform like new ones, the building envelope holds the key to fulfilling the full performance potential of existing buildings.

New building codes are emphasizing these very same strategies. The International Energy Conservation Code of 2012 requires air duct tightness, mandatory blower testing, rigid foam on the exterior of houses in cold climate zones, additional wall insulation, increased glazing U-factor and SHGC, and energy-efficient lighting and equipment.

The requirements in the 2012 IECC are not that different from Passive House standards. It is as if a building industry that has long focused on improving energy efficiency in buildings by upgrading the efficiency of the mechanical equipment has had a collective awakening about the central feature of energy-efficient design, having the epiphany that "It's the envelope!" The new codes recognize that the real key in reducing energy use is the building enclosure. To some extent, Passive House principles are already part of building codes in Brussels, Lower Austria, Wels, Antwerp, Bavaria, Bremen, Darmstadt, Frankfurt, Freiburg, Hanover, Heidelberg, Cologne, Leipzig, Kempten, Nuremberg, Munster, Lindenberg, Ulm, Saarland, Rhineland, Walldorf, Luxembourg, Oslo in Norway, Villamediana de Iregua in Spain, and Marin County in California. There are currently further incentives to utilize Passive House standards in San Francisco and New York City.

So, whether you look at the increasing number of cities adopting these building codes or you look at the building codes themselves, Passive House building science is building momentum. Passive

House is the main strategy to reach Architecture 2030 Challenge Goals and Zero-Energy, precisely because its implementation addresses cost-effectiveness in an industry historically cost-conscious and resistant to change.

Conclusion

SEVEN CRITICAL (YET SOMETIMES FORGOTTEN) STEPS FOR HIGH-PERFORMANCE DESIGN AND EXECUTION

1. Early Planning: Passive House is a systemic approach to construction. All team members need to be involved in the project as early as pre-planning. During this time, performance goals for energy consumption and indoor environmental quality are established and team members begin to internalize their roles in the delivery of the performance goals.
2. Building Massing and Orientation: Very early in the project, building-site orientation is chosen to derive maximum benefit from solar gain while minimizing losses from heating and cooling systems. It is also at this time that landscape architects will begin thinking about options for contributing to the overall performance of the building.
3. Envelope Optimization:
   a. Insulation must be continuous and tailored to the climate zone.
   b. The air barrier must also be a continuous layer around the entire envelope.
   c. Window performance must be balanced with the performance of the envelope assembly in order to maximize winter heat gain and minimize summer heat gain. Ideally, building owners prioritize a high solar-heat-gain coefficient (SHGC) to absorb free heat and a low U-value to prevent heat loss during evening hours.
4. Ventilation: In Passive House construction, airtight buildings require the use of heat- and moisture-recovery ventilation systems to continuously introduce outdoor air ventilation.

5. Systems and Envelope Commissioning: Systems commissioning is routinely considered, but project teams looking to reduce their building-performance risks and financial risks should engage an experienced building-envelope commissioning agent to ensure that the continuous insulation and continuous air barrier are constructed properly.

6. Measurement and Verification: To ensure that high-performance building goals and targets are met, building owners should implement feedback loops to visualize building performance based on the key performance indicators set during planning. An integrated performance dashboard will compare building performance trends in operations with performance metrics set in early planning. This type of dynamic display enables building owners to gain confidence in investing in existing buildings. (For details, see chapter 7.)

7. Communication among Team Members: The life cycle of a traditional construction process is typically a hand-off from the major phases of design through construction and into operations. Even in the best projects, the design and construction phase is almost always disconnected from operations. If you need proof, just ask your facilities manager to show you a project's Owners Performance Requirements from the building design.

Project Development Homework

✔ Consider the envelope conditions of your target project. Specifically, focus on the thermal barrier, air barrier and air leakage, and thermal bridging.

✔ What changes would you instinctively make to improve the overall performance of the envelope in terms of energy consumption and indoor air quality?

✔ Try to calculate the impact of those improvements on long-term operating costs, human health, and the productivity of your building occupants.

Chapter 6:

How Realistic Is Zero-Energy for an Old Building?

No problem can be solved until it is reduced to some simple form. The changing of a vague difficulty into a specific, concrete form is a very essential element in thinking.
—J. P. Morgan

M ANY PEOPLE THINK ZERO-ENERGY goals for new buildings are cost-prohibitive. We will show you that Zero-Energy goals can be achieved within a reasonable budget for new buildings and they are also possible when restoring and renovating existing buildings. To accomplish these goals within a reasonable budget, all that is required is to create a project design and develop construction processes that are guided by technology with a firm commitment to building science. Technology and building science give owners the basis from which to make data-driven decisions. What is behind Zero-Energy is the knowledge that it's possible for existing buildings to be self-sufficient. To reach self-sufficiency, the industry is transitioning to low-carbon expectations, which requires energy-efficient solutions for the existing building stock as well as for all new buildings. To this end, we know that "it takes between ten and eighty years for a new building that is 30 percent more efficient than an average-performing existing building to overcome, through efficient operations, the negative climate-change impacts related to the construction process."[1]

While the United States constructs roughly 2 percent of new commercial floor space each year, most opportunities to improve efficiency over the next several decades will be in the existing building stock. Most of these structures are constrained by old equipment, aging infrastructure, and inadequate operations resources. Improving the efficiency of existing buildings represents a high-volume, low-cost approach to reducing energy use and greenhouse-gas emissions.

In developed economies, at least half of the buildings that will be in use in 2050 have already been built. According to a recent survey by the US Energy Information Agency, 72 percent of floor stock in the United States (46 billion square feet) belongs to buildings over twenty years old.[2] We simply don't have the resources to raze all poorly performing buildings and start over.

As history has taught us, breakthrough solutions to complex problems like Zero-Energy for the built environment typically require collaboration between governments, for-profit companies, universities, industry associations, and not-for-profit organizations. We have seen governments step in across developed countries with policies intended to incentivize energy self-sufficiency. While not broadly known as a global leader in sustainability, Pittsburgh, Pennsylvania, has some of the most aggressive policies regarding energy use in the built environment. In 2016, Pittsburgh ratified the Energy Disclosure Law requiring buildings within the city to disclose to the public the energy use of buildings, based on energy-use intensity (EUI). Around the same time, the City commissioned the development of the "Pittsburgh p4 Performance Measures" (People, Planet, Place, and Performance), which details how building owners and developers should think about investment in real estate projects located in Pittsburgh. The Energy Disclosure Law and the p4 Performance Measures support the City of Pittsburgh's 2030 Challenge Goals of carbon-neutral new buildings by 2030 and a 50 percent reduction in energy use in existing buildings by 2030.

New York City outlined an aggressive vision for reducing carbon emissions by 80 percent by 2050 in its "One City: Built to Last" plan.[3] This document suggests that improving the envelopes of buildings by using Passive House strategies is the path to reaching those goals. In their effectiveness in reducing carbon emissions, these measures are considered second only to changing from coal-fired power plants to cleaner sources for the generation of electrical energy. The result has been an explosion of new Passive House and near-zero-energy buildings in the city. As we mentioned in chapter 1, part of New York's One City: Built to Last plan is recently enacted legislation requiring owners of large buildings to reduce their contribution to carbon emissions by 40 percent in the next eleven years or risk significant fines.

Yes, these are very aggressive goals, especially in the United States, and that might be a problem if they didn't make so much sense for the planet and society as a whole. Many owners who are inexperienced in building efficiency look to sustainability certification programs for guidance (see box 1-1 and fig. 3-1). But it is worth mentioning here that sustainability certification programs do not always align with the goals set by building owners. Existing buildings have complexities and constraints that new buildings do not have, so an evidence-based approach to building efficiency is necessary to remove the uncertainty of Zero-Energy goals. With an evidence-based metrics process, building owners can "dial in" their energy-efficiency solutions while balancing the real constraints of project budgets and annual operating expenses.

So, do clear and convincing solutions exist to transform existing buildings to Zero-Energy? While older buildings do use a great deal of energy, the answer is YES—existing buildings can indeed be transformed to Zero-Energy. The key to realizing a built environment that is comfortable, efficient, and cost-effective is to unlock the vast potential for energy efficiency in existing buildings. While it is possible to upgrade buildings step by step, whole-building retrofits create a holistic set of opportunities that piecemeal approaches do not allow.

Once building owners commit to transforming existing buildings to Zero-Energy, the next step is to define the language without which measurable results are difficult, if not impossible, to achieve. The metrics used to establish goals in performance-based contracting specifications affect how buildings are designed to achieve the goal. High-performance building projects often reference the concept of "Zero-Energy buildings." Critically, the question becomes "How do you define Zero-Energy buildings"?

Set Goals of Zero

The National Renewable Energy Laboratory (NREL) defines Zero-Energy buildings as "residential or commercial buildings with greatly reduced energy needs through efficiency gains such that the balance of energy needs can be supplied with renewable technologies."[4] NREL's white paper "Zero-Energy Buildings: A Critical Look at the Definition" explains why a clear and measurable goal is needed in Zero-Energy projects and offers several definitions of Zero-Energy buildings.[5] Which definition may be appropriate depends upon project goals and the values of the design team and owner. Here are four well-documented definitions:

1. Zero-Energy Site Energy: a site Zero-Energy building that produces at least as much energy as it uses in a year, when accounted for at the site. "A Zero-Energy building that produces at least as much energy as it uses on site."
2. Zero-Energy Source Energy: a Zero-Energy building that produces at least as much energy as it uses in a year, when accounted for at the source. "A Zero-Energy building that produces at least as much energy as is used in a year that is produced and delivered to its site."

3. Zero-Energy Energy Costs: a Zero-Energy building where the amount of money the utility pays the owner for the energy the building exports to the grid is at least equal to the amount the owner pays the utility for the energy services over the year.
4. Zero Energy Emissions: a Zero-Energy building that produces at least as much emissions-free renewable energy as it uses from emissions-producing energy sources.

The NREL study reveals the significant differences among various definitions. Depending upon the goals set by the owner, project teams will implement different project strategies that may result in significantly different energy-utilization outcomes. Further, each goal may utilize different energy-conservation measures requiring various renewable-energy options. The NREL research study is a must-read for any team member engaged in performance based contracting.

Building owners and developers in Rust Belt cities such as Pittsburgh are keen to find paths to reach Zero-Energy. With an abundance of older buildings, some of the city's most intensive users of energy see the strategic nature of the problem. The more successful companies are, the greater the challenge to reduce their consumption of energy. Setting a goal of Zero-Energy requires a strategic imbalance of sorts to deal with the organic increase in energy-use intensity caused by the additional people and process load that naturally comes from success. Reducing whole-building consumption of energy while simultaneously adding loads requires a strategy. Existing buildings pose a special challenge because, many times, the existing buildings were built for purposes significantly different from their current use.

Evidence of What Is Possible

Originally constructed in 1918, the Wayne Aspinall Federal Building was renovated in 2013 with a typical goal of modernization as well as an aspirational goal of becoming the first Zero-Energy building listed on the National Register of Historic Places. This is not only important for the significant improvements available to historic places, but also because the federal government is the largest building owner in the United States. Prior presidents and the General Services Agency have mandated the measurement of GHG emissions and reductions in building CO_2^e footprints.[6]

The 2013 renovations successfully converted the building into a high-performance leader of energy efficiency and sustainability, while preserving its original character. Net Zero-Energy objectives are met through a combination of high-performance, energy-efficient materials and systems, and on-site renewable-energy generation. This building is now 50 percent more energy-efficient than a typical office building. On-site renewable-energy generation is intended to produce 100 percent of the facility's energy needs throughout the year.[7]

There are also growing examples of Zero-Energy case studies for residential buildings. The Davis House in Somerville, Massachusetts, is but one. The renovation transformed this building from an energy guzzler to a resilient and healthy two-family home, designed to use less energy than it produces. Proper planning resulted in high efficiency of the building envelope, heating systems, cooling systems, and ventilation systems. Solar panels provide the energy needed to run this two-family home. This is important because existing homes outnumber new housing nearly 300 to 1. There are 130 million housing units that already exist, and 80 percent of them will still be here in 2050. The goal for the Davis House was to reduce the energy use and increase the usability of the building within the same footprint by focusing on the envelope, retaining 60 percent more solar gain on the south side of the house for passive solar heat, adding windows to the south side, removing windows on the north side of the structure, and by providing solar shading.[2]

The number of Zero-Energy homes is growing. One of the authors of this book did most of their writing within their own Zero-Energy home—a building producing enough energy to power all the annual needs in the home and an electric vehicle with a cost for this renewable energy at 40 percent less than fossil-fuel-based electricity provided by the local utility company.

While we next draw from a new-construction example, there is good reason for doing so as the same concepts applied to the Bullitt Center in Seattle, Washington, can also be applied to existing buildings. The Bullitt Center is a highly publicized project that received Living Building certification. On their website, CEO Denis Hayes reports:

> The six-story, 52,000 SF Bullitt Center in Seattle, Washington [...] is one of the most energy-efficient commercial buildings in the world—a high-performance prototype setting innovative standards for sustainable design and construction, demonstrating that it is possible to create a commercially viable building with essentially no environmental footprint.

The purposeful use of a tight envelope and integrated design focused on energy efficiency enables the building's operational energy use to be net positive, and the Bullitt Center projects it to remain that way.

This building demonstrates how a building owner, an integrated design and construction team, supportive regulatory agents, and progressive financial partners can come together with a common purpose to achieve extraordinary results. This building serves as an inspiration for the creation of the next generation of high-performance buildings. It illustrates some of the critical elements of both the integrated design process that guided the building's development and construction and also the integrated design systems employed in its operations.

The Bullitt Center has exceeded expectations for thermal comfort and daylighting as well as for energy use during its first year of operation. Thanks to the envelope, the building is warm and draft-free in the winter, cool and comfortable in the summer, and beautifully daylit year around. Occupants of the building express a high level of satisfaction with the quality and comfort of the indoor environment.

The project team's engineers established two benchmarks to measure the building's energy performance against. First is the EUI (energy use intensity) for an average office building in Seattle (Energy Star score = 50), which has an EUI of about 72 kBtu/sf/yr. The second is an approximation for a 2009 Seattle Code building built on this site, which has an EUI of about 42 kBtu/sf/yr. The target EUI for the building was 16.1 kBtu/sf/yr. From May 2013 through April 2014, the first 12 months of occupancy, the building's EUI was 9.4 kBtu/sf/yr, about 41 percent better than predicted performance and 77 percent better than a 2009 Seattle Code minimum building.

Occupancy accounts for part of the Bullitt Center's exceptional energy performance. On average the building was occupied at half of its design occupancy during the first year. Since about half of the building's predicted energy use is from "activity loads" directly tied to the number of people using the building, the corrected target for the building's energy use is about 12.3 kBtu/sf/yr. Experts are still working to this day to understand how this building uses energy and how its performance is exceeding predictions. Whole-building energy and power use, as well as energy production data, have been collected since the photovoltaic (PV) power-production plant went online and began supplying the building with energy in early 2013. But while every circuit in the building can be monitored, validating the end-use, circuit-by-circuit data is still a work in process.

It's important to note that the Bullitt Center achieved extraordinary results on every level of operations, but its design did not emphasize a high-performance building envelope. The benefits of a high-performing envelope make Bullitt Center type results easier and less expensive to achieve. See figure 6-1 for generalizable examples from the Bullitt Center for reducing any building's EUI.

Whether an owner is contemplating taking new construction to

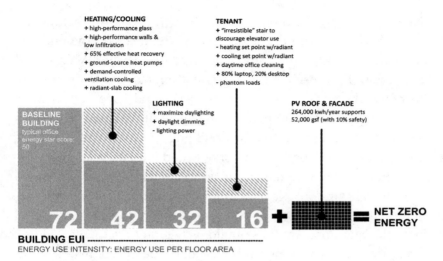

Figure 6-1. The Path to Net-Zero Energy (used with permission from Rob Peña, author of "Living Proof: The Bullitt Center").

Zero-Energy or retrofitting an existing building, the Natural Order of Sustainability remains the same: passive first, active second, renewables last. Only after we are confident that we have taken an existing building to its theoretical optimal energy performance does it make sense to develop plans for renewable energy.

One of the first Zero-Energy buildings ever completed in the United States was completed at the Center for Environmental Studies at Oberlin College by Bill McDonough and Partners. The Oberlin building attempted to achieve Zero-Energy status through building orientation, geothermal wells, an earth berm on the north side, energy-efficient equipment, and the installation of an enormous PV array costing $420,000 at the time. The 45-kW PV array covered the entire rooftop. This project was built prior to the advancement of energy modeling and simulation technologies. Despite Oberlin's ambitious and admirable goals, they still fell short of providing all of the energy that they needed. They then built a second 100-kW PV array that covered a large parking lot adjacent to the building.

Now they are able to produce enough energy to cover their needs. There was no way to fit the entire array required on the building itself without using the additional space on the site. What would the results have been if there was not enough space on the site to house the extra panels? The project would have had to settle for making less than 50 percent of its goal. It's reasonable to wonder, had Oberlin optimized their building envelope first, whether a smaller and less expensive PV array would have been enough.

In Europe there are numerous buildings with PV integrated into the vertical skins of large tower projects in urban areas. These offer some relief to the issue of real estate, but replacement cost still begs the question: Are those buildings going to get new cladding every twenty years? That kind of capital expenditure may work in Europe, but we don't have the financial structures in place in the United States to make that strategy a reality.

There are all kinds of existing buildings that will require retrofitting to reach any level of energy efficiency. Some historic cultural treasures will never be retrofitted because insulating and air-sealing them would ultimately destroy their character. Other buildings cannot be retrofitted because of a lack of money. If we are going to hit the global emissions target by 2050, then existing buildings must be retrofitted to produce as much or more energy than they use. As we develop more and more knowledge about ways to retrofit existing buildings, the costs will continue to come down.

We introduced the East Liberty Presbyterian Church (ELPC) cathedral project in chapter 1. You may recall that the outcome of the project resulted in a final design that met the project budget, realized the energy goals of 30 percent less usage that its pre-construction performance, and created superior indoor air quality throughout the entire cathedral. ELPC completed construction at the end of 2018, and the current energy trends meet the predicted performance targets. It is the first cathedral in the world to achieve RESET Air Certification for Interiors. While this is not yet a Zero-Energy building, the approach taken by the project team sets ELPC up beautifully for a future capital campaign dedicated to renewable energy. By reaching the church's theoretical optimum performance in terms of energy using passive and active strategies, ELPC has in place the most cost-effective path possible to reach Zero-Energy using renewable strategies.

It is noteworthy to discuss ELPC's Natural Order of Sustainability strategies; not because ELPC was able to dramatically improve its thermal and air barrier, but because they could not. As a historic building with limestone cladding, stone and plaster interior walls, and original single-pane, aluminum-framed windows, passive retrofit strategies were challenging. At first, the project team refused to even consider passive strategies. The first schematic design resulted with construction costs double the intended budget, unaffordable annual operational expenses, and too many unknowns regarding performance goals.

After the project team went back to the drawing board, we turned our attention to building science. What could we learn, using technology, about the potential of the building? Through whole-building performance modeling we learned that ELPC, designed by renowned architect Ralph Adams Cram, was designed in 1930 to be a high-performance building. Our motto became "restore and improve the building systems as originally designed." We also used the building's thermal mass to our advantage. Rather than trying to restrict thermal comfort in a close band of 68 to 72 degrees, we expanded the range and looked for dehumidification and air-movement strategies.

The project had many passive strategies, despite the limitations of historic design, including the following:

1. Restore the aluminum windows to repair broken frames, glass, and operators.
2. Remove aluminum oxidation scale from window frames and add weathertight gaskets to minimize air infiltration.
3. Remove and replace window caulking.

4. Install new glass-door airlocks with weathertight sealant at the main entrances.
5. Seal off selected interior rooms to receive air-conditioning.
6. Add weather stripping to all the existing outside entrance doors.
7. Air-seal all mechanical penetrations throughout the building envelope.

Our active strategies included the following:

1. Decouple the HVAC solutions as originally planned.
   a. Add a dedicated outdoor air system (DOAS) system to provide continuous, filtered, and dehumidified fresh air.
   b. The DOAS system has in-line blower-coil unit cassettes (BCU) to provide air cooling in selected rooms.
   c. Corridors, restrooms, and selected interior rooms receive ventilation air, but not air cooling, from the DOAS.
2. Install new ceiling paddle-fan units in all occupied rooms.
3. Make sure all windows are operational.
4. Add a new building management system and combine all active systems in order to optimize schedules, sequences, and set points for heating, cooling, and ventilation.
5. Put the boiler-fed steam radiators on a whole-building building-management system for setback and zone control.
6. Schedule nighttime purge ventilation to take advantage of free cooling.
7. Upgrade the light fixtures to LED with occupancy sensors.
8. Ensure that all new appliances and equipment are Energy Star certified.
9. Add an energy-management platform to measure and monitor whole-building performance of energy and indoor air quality using primary-source digital utility meters, indoor air-quality monitors, a weather station, an operational whole-building performance model, and a building management system.

Conclusion

By respecting building science and efficiency strategies, building owners are now beginning to understand that there is a powerful relationship between building energy performance, on-site energy generation, and affordability. By reducing demand first, the original plant size for on-site generation will likely become smaller, easier to implement, and much less costly. The same is true when systems already in use reach the end of their lives. Lastly, it's important to remember, when renewable systems break or are interrupted in a highly efficient building, the negative impacts on operating costs are far less than they would be on less efficient buildings.

Meaningful progress toward low-energy, high indoor air quality in existing buildings is not achieved by using single, disconnected strategies like LED light fixtures, variable-frequency drive motors, or photovoltaic arrays. Superficial changes to existing buildings will not transform the built environment in a manner that is required to reverse climate change. The silver bullet, we believe, is the process used to design and construct buildings. Since the 1970s, we have learned that in order to achieve aspirational results we must look at buildings as living organisms. Natural processes and designs set the standards to which we aspire. The Natural Order of Sustainability provides an organic and affordable pathway to reach Zero-Energy without paying a significant financial premium or sacrificing thermal comfort. The path to true reduction in energy use in the built environment requires vision, patience, and a commitment to energy master-planning.

Lastly, and most logically, the byword for sustainability in the built environment must become "the cheapest form of energy is the energy never used."

Key Questions for Your Project

As we strive for building improvement, we must ask: How far can we push a project to get to Zero-Energy as we tunnel through traditional barriers and diminishing returns for incremental changes? Does our next decision positively impact climate change? Are we, in the AEC community, doing enough to reverse energy consumption trends? What we do know is that, if we continue with a business-as-usual model, building energy consumption and related GHG emissions will continue to contribute to climate change and continue to threaten human health and productivity.[9] What if all existing buildings could get to Zero-Energy?

When we look at buildings, we see complex objects operating dynamically. Changes to individual energy-conservation measures typically do not result in directly proportional reductions to total building energy consumption. However, changes to groups of energy-conservation measures typically have a compounding impact on reduction of total energy consumption. Where do we start? We start by setting performance metrics early in the design process and then measuring the building's performance against those goals in operations. Only at that point can we know the full costs and benefits of taking any building to zero-energy performance.

Every existing building has the potential to reach Zero-Energy, provided project teams make the right choices in the proper sequence. We encourage project teams to embrace technology early in a project to allow enough time to simulate building performance prior to making any decisions on source/supply of energy. By taking the steps outlined in earlier chapters, project teams will have enough confidence in costs and performance to quickly see the high-performance building design options that make high-performance building operations possible.

Project Development Homework
   ✔ Break your Zero-Energy opportunity down into subsystems.
      • Which subsystems present the best business case for change?
   ✔ Where is a logical place to start for your building?
   ✔ Can you create a master plan that can be implemented in

phases in order to take advantage of natural triggers of life cycle, deferred maintenance, renovations, and other conditions to reach Zero-Energy over a period of time?

✔ If you cannot get to zero, how low can you go?

Chapter 7:

Operating Buildings for Maximum Benefit

If you can't measure it, you can't improve it.
— Peter Drucker

"DON'T JUST TELL ME, SHOW ME" is a big part of this chapter as we focus on reliable measurement and verification systems that provide feedback on the key performance indicators defined in the building owner's goals and targets. Reliable and cost-effective measurement and verification systems connect all the critical elements of smart buildings, equipment, and systems. Understanding the minimum infrastructure required for a reliable measurement and verification platform enables building owners to layer in new sensors and meters as technology develops. In general, technology changes the game in terms of how energy efficiency and indoor environmental quality are delivered and verified. Building performance simulation and smart building infrastructure allow owners to balance building performance against first costs and long-term operating costs. With the help of a technology platform, owners will know what is possible prior to spending construction dollars. Remember, so far we've shown you how building science, done properly, enables any existing building to perform like a new one. In this chapter, we will show you how technology will increase your confidence to invest in existing buildings by increasing the knowns and decreasing the unknowns of any building project.

The most recent breakthrough in data analysis for the built environment is that it's now possible to embed dynamic simulated-performance targets next to trended performance. We call this an integrated sustainability dashboard. Simulated targets from an operational energy model, are displayed next to trended performance from the actual performance of the building. This integration provides transparency in addressing "How was the building originally designed to perform?" and "What will performance look like if we invest in building improvements?" The norm today is to wait for twelve months of recorded trended performance to determine the success or failure of previous energy-conservation measures. Using an integrated sustainability dashboard, it's possible to know immediately if the measures taken have delivered the expected results.

There is a relatively small but growing number of forward-thinking people already assessing building performance during operations for energy use and IEQ against targets set in planning. Later in the chapter, we explain how monitoring-based commissioning works and why connected meters and sensors should remain with the building over its lifetime and be expected to adapt to emerging technology. As we end the chapter, you will also see how little it actually costs to invest in basic technological monitoring infrastructure for new and/or existing buildings.

Peter Drucker, a well-known management scholar, consultant, and educator famously said, "If you can't measure something and know the results, how can you possibly expect to improve it?" For example, it's nearly impossible to lose weight without stepping on a scale to measure your results. If you don't

Figure 7-1. Vehicle Dashboard.

monitor your weight, you have no idea if you are succeeding or not. If you are trying to improve your golf game, but never keep score, you don't know if you're getting better. Similarly, if you don't know your building's performance metrics, then you can't possibly manage or improve them.

For those of us who still own an automobile, think about your vehicle. What is your current gas level? How fast are you traveling? The key performance indicators for the automobile have been clearly defined. They are displayed in real time and with context (see fig. 7-1). People from ages sixteen to ninety can sit in a car for the first time and understand the information displayed on a vehicle's dashboard.

Building performance should be no different. People spend 93 percent of their time indoors. Think about the building you are in right now. What is the energy-use intensity of the building? What is the air-quality level in the building as you read this book? If you are like the other 99 percent of building occupants, building owners, architects, engineers, and contractors in the world, you don't know.

Benefits and Limitations of Building Management Systems

Current trends in the market include repurposing the building management systems (BMSs), sometimes referred to as building automatic temperature control (ATC) systems into building performance dashboards. At first glance, it seems logical that filling the data void with a system would help building owners who are struggling to understand their building's performance. BMSs, by their design, have access to temperature, humidity, CO_2 levels, fan operations, schedules, set-points, and building system sequences. Add to this the reality that building owners are reluctant to purchase and install another system in their building because it will require additional maintenance, training, and expense. Many building owners are heavily invested in building management systems. So the argument to repurpose the BMS into an energy management system makes sense on the surface.

Unfortunately, the realities of repurposing BMSs exposes many limitations, such as security, ease of use, system integration, and data access across a number of platforms. Common BMS shortcomings include limited data-archiving capabilities (time, granularity, flexibility), limited user-friendliness in accessing, visualizing, and sharing the data, limited enhancements to proprietary legacy systems, and limited capabilities for integrating with other systems (i.e., power meters, HVAC equipment, lighting, security, fire alarm, and so on).

A smart-building approach differs from a traditional BMS approach in which engineering, facilities, and enterprise systems are connected via their own discrete infrastructures. A smart building facilitates connection of any system over a common communications infrastructure (cabling, network infrastructure) using industry-standard open protocols and application program interfaces (APIs), allowing data to be shared and analyzed to provide cause and effect mapping between systems.[1]

Security is perhaps the biggest argument against using a BMS as an integrated sustainability dashboard. Owners and operators need to see performance data, in specific and aggregated form, in real time.

Integrated sustainability dashboards need to be "open" and accessible to everyone from building occupants, facilities managers, building owners, and future design teams, to other systems and equipment within the building. Opening a BMS to people such as external facilities managers and independent systems and equipment creates a security risk, enabling potential access to mechanical equipment and automated temperature-control systems. This security risk is real and leads to a privatization of data that is counterproductive to installing a comprehensive integrated sustainability dashboard. Operators of BMSs correctly protect their systems with firewalls and security measures that restrict access to data. As nontechnical stakeholders desire access to the building performance data, the only access point is typically through the BMS. When the BMS becomes the gatekeeper to a building owner's performance data, it is understandable that owners feel frustrated and are unwilling to pay for access to information they feel is already theirs.

As an example, we offer the well-known Target breach. In managing cybersecurity risk, many real estate owners are cautious about adopting next-generation technologies because of the real and perceived cybersecurity risk associated with building systems that are managed in the cloud or linked to other critical business systems behind a company firewall. After the massive 2014 hack, Target's systems breach was traced back to their HVAC vendor, and, after hackers proved they could hold building automation systems hostage in the middle of winter in Finland in 2016, company chief information officers have been reluctant to approve new technologies that could provide a back-door route into other critical systems.[2]

The next difficulty is in the complexity of the BMS. Many automation and control systems are complex and require significant training. Integrated sustainability dashboards should be simple and intuitive. No one sitting in a car needs more than five minutes of orientation to understand the dashboard. Successful building performance dashboards should have the same level of intuitive comprehension and user ease. Nevertheless, too many BMSs struggle with balancing their core competencies of controlling temperature and humidity with an additional requirement of feeding data into a user interface in the form of a dashboard.

Dashboards

A well-thought-out building performance dashboard should be intuitive to all users. The data should be easily organized by area and/or use with easy transitions between energy consumption, energy generation, air quality, light levels, sound levels, occupancy levels, and other key performance indicators as defined by the building owner. The best building performance dashboards provide visual context for the data—for example, how the building is designed to operate, what the performance levels of similar buildings are, or how the data compares with performance targets. If you can't tell how your building measures against key performance indicators easily and quickly, then more work is needed to develop a better summary of performance. Instead of simply displaying trend data directly from meters and sensors, consider integrating trend data in a visual context with dynamic targets as shown in figure 7-2. In its simplest form, static threshold targets may be used from sustainability certification programs. Better yet, dynamic targets from the energy model used in planning can be contrasted against trended operational performance. For example, if an owner decides that CO_2 levels exceeding 800 ppm are unacceptable, a static threshold target can be displayed at a flat 800-ppm level (see fig. 7-2). Static threshold targets are acceptable for some limited key performance indicators, like particulate matter (2.5 and 10 microns), total volatile organic compounds, light intensity, and sound levels.

However, when considering all forms of energy, temperature, relative humidity, and CO_2, static targets are relatively useless. Dynamic targets that adjust for occupancy, seasonality, weather, and other key performance indicators with a great deal of load variability provide a level of control and understanding that is nec-

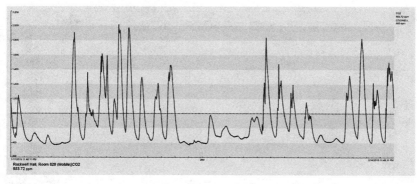

Figure 7-2. Trending CO₂ Data with Thresholds, Rockwell Hall, Duquesne University.

essary for operating high-performance buildings. Dynamic targets from an operational whole-building energy model can be displayed in the same time interval as trended data. Visual representation of the performance intended from a building helps us understand how the building should perform. When existing-building owners can see data in dynamic dashboards, they gain confidence in future investments in the building. Much like dieters tracking their progress against the bathroom scale every morning, building owners want to know if they are winning or losing. Building performance dashboards can answer that simple question.

Fortunately, dashboards are evolving to incorporate new data that facilitates better ad hoc analysis within buildings, just as they are within cars. For example, this Tesla dashboard (fig. 7-3) quickly shows the driver the current battery level (there are 156 miles left on the battery) safety features, speed limits, energy consumption, range, and provides one-touch access to other operable aspects of the vehicle.

Building owners and occupants are demanding real-time answers that can be accessed on their smartwatches, smartphones, and tablets. Further, many building owners and most sustainability certification programs understand the value of empowering building occupants and stakeholders, and therefore they take steps to provide them with real-time access to building performance information

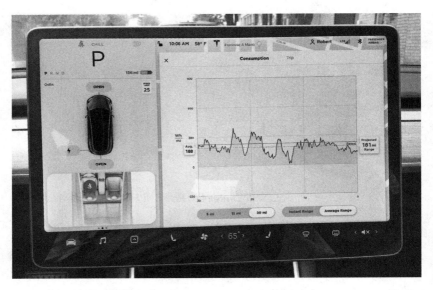

Figure 7-3. Tesla Dashboard.

(see fig. 7-4a-d). Owners want dashboards to be scalable across their buildings in a cost-effective manner. They want the interface to be "open" and to be ready for future Internet of Things (IoT)–enabled sensors, meters, and other technologies. Many argue that the IoT can be used to create new models of engagement if device networks can be open to user control and interoperable platforms. BMSs, by their nature, cannot be open without significant security risks.

A truly open measurement and verification platform can create visualizations using "single pane of glass" solutions. This kind of solution presents a holistic view of data from multiple sources throughout the building. A unified display allows stakeholders to view and control multiple systems, regardless of manufacturer, from a single visual interface. Users can glance at key performance indicators in numeric and graphical formats. Facilities managers can monitor specific devices for real-time data and explicit detail. Decision makers can understand from a top-level view the performance of their buildings.

Figure 7-4a. Duquesne University: Display of all buildings in a portfolio with whole-building performance metrics rolled up into a campus-wide building performance metric.

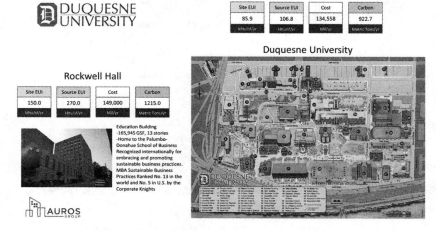

Figure 7-4b. Duquesne University: Display of whole-building performance for a single building rolled up into a campus-wide performance portfolio.

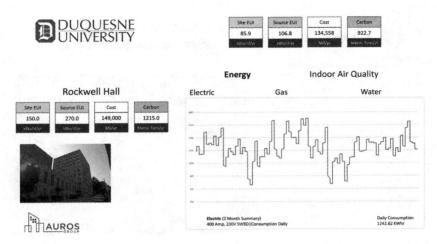

Figure 7-4c. Duquesne University: Display of whole-building performance for energy consumption with a dynamic single-building target rolled up into a campus-wide performance portfolio.

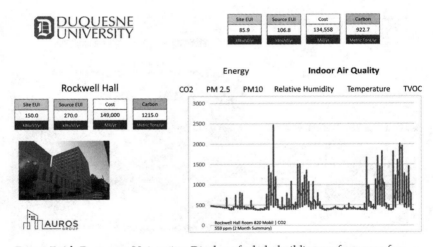

Figure 7-4d. Duquesne University: Display of whole-building performance for indoor air quality with a static single-building target rolled up into a campus-wide performance portfolio.

In short, an integrated sustainability dashboard is not a BMS, nor should it be. A BMS trapped behind a firewall without easy access to data is just not compatible with the macro trends of sensors and meters using Internet of Things (IoT) connectivity.

Internet of Things–Connecting Systems

The term "Internet of Things" (IoT) was coined by Kevin Ashton of Procter & Gamble in 1999, though he prefers the phrase "Internet for things." At the time, he viewed radio-frequency identification (RFID) as essential to the IoT, as it would allow computers to manage all individual things. IoT is the network of physical devices, vehicles, home appliances, and other items embedded with electronics, software, sensors, actuators, and connectivity, which together enable these things to connect and exchange data, creating opportunities for more direct integration of the physical world into computer-based systems, resulting in efficiency improvements, economic benefits, and reduced human intervention. The number of IoT devices increased 31 percent year-over-year to 8.4 billion in 2017, and it is estimated that there will be 30 billion IoT devices by 2020. The global market value of IoT is projected to reach $7.1 trillion by 2020.[3] The market for the IoT is continuing to grow at a phenomenal pace. IHS Markit forecasts that the IoT market will grow from an installed base of 15.4 billion devices in 2015 to 75.4 billion devices in 2025.[4] Other market research firms are releasing similarly staggering statistics, and while estimates vary, all parties agree: network-connected devices and their capabilities are, and will continue to be, disruptive forces in the way people and businesses interact.

If building managers don't plan for this network integration, every new technology will have to create its own independent network within buildings, which is precisely what we see in buildings today. For example, many buildings have a telephone system network, information-technology network, building security and

CCTV network, fire alarm network, building automation control network, energy consumption network, and indoor environmental quality network, and so on. This ad hoc, siloed approach to network installation is wasteful, shortsighted, and absurd when the future is about connectivity. Truly smart buildings address this directly through a single, interconnected network infrastructure.

Significant numbers of meter and sensor devices in the built environment already integrate via the Internet, which provide the potential for them to communicate and be centrally managed via cloud-based interfaces. These devices monitor air and water quality, light and sound intensity, atmospheric or soil conditions, movements of occupants, and a growing list of other factors. Creating an open environment of meters and sensors within a data network that is robust, durable, safe, portable, and scalable is the goal. Integration of BMS and other discrete systems via an IoT-based sustainability dashboard is logical and necessary to drive building performance. Unfortunately, most owners are unaware of the drawbacks of placing their IoT-based integrated sustainability dashboards under a proprietary, that is, a "closed" BMS. However, there are ways to build upon the BMS platform and achieve integrated performance without having to start over and forgo the investment made in the existing BMS.

A properly planned IoT-based integrated sustainability dashboard enables real-time monitoring, which facilitates reducing energy consumption, improving indoor environmental conditions, and monitoring occupant behaviors. The integrated sustainability dashboard should mirror the key performance indicators based on the goals set in early planning and the information covered in the earlier chapters of this book. The IoT integrated sustainability dashboard becomes the basis of an adaptable infrastructure that can integrate plug-and-play devices and other future developments.

The development of an IoT-based integrated sustainability dashboard requires a platform for interconnected devices. With nearly half a million installations worldwide, Niagara is quickly becoming the operating system of the Internet of Things in the built environment.[5] Its open API, open-distribution business model, and open-protocol support provides the freedom to scale up and down with meters and sensors to address the key performance indicators related to the building. A Niagara platform connects and controls devices while normalizing, visualizing, and analyzing data from nearly anywhere or anything. With its compact, embedded IoT controller and server platforms, Niagara connects multiple and diverse devices and subsystems. With Internet connectivity and portability, Niagara integrates control and compatibility with numerous commercially available "front-end" dashboards. It streams data-rich graphical displays to a standard Web browser via an Ethernet cable or wireless LAN, or remotely over the Internet. The Niagara platform is flexible and expandable within a building and across buildings. In larger facilities requiring multi-building applications and large-scale control systems, a Niagara platform can control and aggregate information, including real-time data, history, and alarms, to create a single, unified application.

Once the basic infrastructure is in place, the buildout and expandability of the integrated sustainability dashboard should depend upon the key performance indicators established for the building. It is critical to monitor and measure the goals established by the building owner in order to create feedback loops to determine the success or failure of the retrofit project. At a minimum, you will want to consider and review energy and water consumption, indoor air quality, and indoor environmental quality. However, finding the correct sensors or meters to meet the building owner's key performance indicators follows the same logic and process as the basic infrastructure. Sensors and meters should be able to "plug into" the integrated sustainability dashboard, which is an important consideration for future expansion of the system. At this time in the evolution of building performance, we don't know what we don't know. But that does not mean that we cannot create infrastructure that allows for plug-and-play integration of multiparameter IAQ (indoor air quality) sensors, wireless geo-fences, biometrics, and other sensors and monitors being developed daily to support occupant-focused key performance indicators. Wearable devices enable occupants to interface with and engage building systems, the dissemination of data, and emerging opportunities for integrated management solutions involving energy consumption and indoor air and environmental quality.

Smart Building Infrastructure

The following components constitute a basic building-wide Smart Building Infrastructure plan (fig. 7-5 shows many of the main components of this infrastructure):

1. Niagara platform interconnection devices (JACE).
2. Meters for electrical, gas, water, and other primary energy sources.
3. Indoor air-quality monitors with a minimum of five (5) parameters for particulate matter (PM2.5 & PM10), carbon dioxide (CO_2), total volatile organic compounds (TVOC), air temperature, and relative humidity, with the capability to be expandable to include radon, ozone (O_3), and other air-quality factors.
4. Indoor environmental quality sensors for light density, sound levels, occupancy quantities, and so on.
5. Weather station with a minimum of six (6) parameters for air pressure, temperature, humidity, rainfall, wind speed, and wind direction.
6. Outdoor air-quality station with a minimum of six (6) parameters for temperature, relative humidity, pressure, particulate matter (PM2.5 & PM10), carbon dioxide (CO_2), and volatile organic compounds (VOC), while being expandable to the measuring of ammonia (NH_3), sulfur dioxide (SO_2), nitrogen dioxide (NO_2), nitrogen oxide (NOx), ozone (O_3), hydrogen sulfide (H_2S), non-methane hydrocarbons (NMHC), lead (Pb), and so on.
7. Operational whole-building sustainability model.

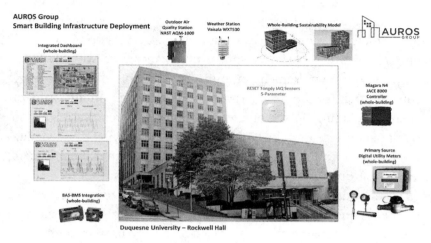

Figure 7-5. Smart-Building Infrastructure.

Building owners will need to select meters and sensors that measure and verify metrics that are aligned with the key performance indicators identified in the owner's performance requirements. Then building owners and stakeholders have the necessary feedback loops for all important metrics. This infrastructure can shrink and grow to match the future needs of the building owners. Final steps now include securing the data, sharing the data in a transparent and open environment, and visualizing the data across portable platforms. This is the work of the integrated sustainability dashboard.

Connecting Key Performance Indicators

Let's review the strategy to measure and monitor *energy consumption* in real time. This takes a few steps to consider, as it is not as simple as connecting digital meters to a display screen. If our key performance indicator for energy is the Energy Star Portfolio Manager (ESPM), then the dashboard should automatically calculate and display site EUI, source EUI, cost, and greenhouse-gas emission metrics.

Building the ideal platform begins with primary source digital utility meters. Unfortunately, most utility providers do not yet permit access to their primary source utility meter, so we may need to install our own. Remember, our focus is whole-building energy consumption, so we need primary source meters on all forms of energy, including on-site energy generation and renewables at the point of entrance into the building. Simply displaying trended data from the meter does not solve the problem, as dashboard users looking at units of measurements (kilowatt-hours, therms, and so on) cannot determine if whole-building goals are being met.

The second step in the process is to take the trended data and display it in the same metric as the target and goals. For example, if we have site EUI energy goals, then we need to convert our energy consumption from its unit of measure to kBtu/sf/yr, as this is the measurement of EUI. This analytical work is best done using the integrated sustainability dashboard. Gathering reliable real-time trended data is the work of the primary-source utility meters and Niagara platform interconnection devices.

Step three is giving similar context to the indoor air-quality data. Using the same smart building infrastructure, we can easily develop a strategy to measure and monitor indoor air quality in real time. This step is relatively simple, assuming the correct selection and deployment of IAQ sensor equipment. To understand IAQ performance, you need to install a qualified and reliable IAQ sensor and connect it to the Niagara platform. The challenge with this component of the smart building infrastructure is that it involves stakeholder alignment on the quality of sensors, deployment, quantification, locations of the sensors, and trended data management from the sensors.

To enable IAQ data and connectivity, we recommend applying the RESET Air standard (fig. 7-6). The RESET Air standard is a perfect example of leveraging an IoT-based integrated sustainability dashboard. The RESET Air standard is rapidly becoming the de facto industry standard and has been adopted into an international consortium for global monitoring standards, which includes the International Living Future Institute (ILFI), International WELL Building Institute, Fitwell, BREEAM, and Passive House Institute US (PHIUS) standards. (See chapter 1, box 1-1.)

The RESET Air standard is an international performance-based standard and certification program for healthy buildings measured in real time with regenerative ecological social and economic targets (hence "RESET"). Indoor air-quality data is gathered through air monitors that measure particulate matter (PM2.5 and PM10), carbon dioxide (CO_2), total volatile organic compounds (TVOC), temperature, and relative humidity. Results stream to the cloud and can be viewed in real time from any computer or mobile device. As a performance based standard, RESET Air begins with the integration of accredited IAQ sensors and monitors. It sets standards for monitoring performance, installation, calibration, and data reporting. It is hardware-agnostic and continuously tests IAQ monitors for compliance. To conform to the RESET Air standard, monitors must be properly installed and commissioned. Monitors must be calibrated, and network connections must be properly configured to stream real-time data to the cloud for the provision of health and certification analytics. RESET Air is the measurement and verification standard that validates trended data while ensuring the quality of data to help understand human health and productivity issues.

Adding sensors using the same smart building infrastructure becomes as easy as plug-and-play. Building owners interested in a strategy to measure and monitor indoor environmental quality to capture metrics for additional indoor air-quality parameters, light, and sound can leverage the flexibility of their smart building infrastructure. Validating key performance indicators for buildings is relatively simple when considering advancements in sustainability certification program and owners demanding proof of performance. This will become clearer in our analysis and discussion of dashboards in the next section. For example,

PM2.5	TVOC	CO$_2$	Temp	RH	CO
Particulate Matter	Total Volatile Organic Compounds	Carbon Dioxide	Temperature	Relative Humidity	Carbon Monoxide
Acceptable < 35 μg/m³	**Acceptable** < 500 μg/m³	**Acceptable** < 1000 ppm	**Monitored**	**Monitored**	**Acceptable** < 9 ppm
High Performance < 12 μg/m³	High Performance < 400 μg/m³	High Performance < 600 ppm	Although there are no requirements for temperature and humidity under RESET™ Air, both must be monitored given their impact on sensor readings for PM2.5 and TVOC.		CO monitors are only required in spaces with combustion.

Figure 7-6. RESET Indoor Air Quality (IAQ) Performance Targets (used with permission from RESET™)

understanding particulate matter (PM2.5) in a building is more valuable when comparing it to outdoor air-quality conditions. Understanding that outdoor air-quality levels are highly variable, many owners now choose to install local outdoor air-quality sensors instead of using public outdoor air-quality sensors located far from the building.

Visualizing Integrated Sustainability Dashboards

The primary purpose of the integrated sustainability dashboard is to convert data into useful information. Too often, engineers get caught up in the data and feel compelled to show data in a graph just because it has been captured. This may seem too obvious, but consider the last building dashboard you visited and ask yourself a question: Was that building winning or losing?

An integrated sustainability dashboard receives dynamic real-time data from digital utility meters, digital weather stations, indoor air and environmental sensors, lights sensors, occupancy sensors, and other building-monitoring devices. It also receives dynamic simulated data from whole-building performance models (see fig. 7-7 below). The dashboard then displays them together on the same time scale, resulting in Web-based access to real-time trended and simulated information for the life of the building.

An integrated sustainability dashboard may further display other important live data in real time, including Energy Star Portfolio

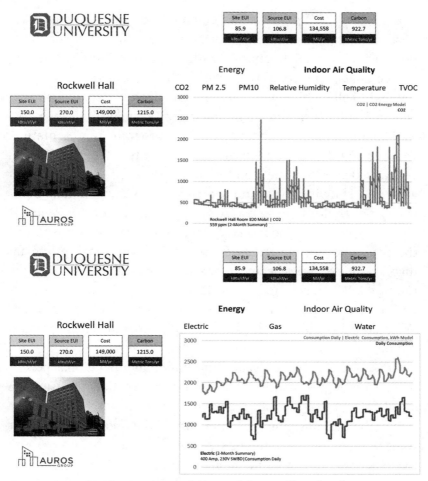

Figure 7-7a and 7-7b . Your Internal Sustainability Dashboard—The Best Way

Manager ratings for utility consumption, CO_2 emission data, predictive utility consumption markers, electric demand, average zone temperature vs. set points, and weather forecast, all to correspond with goals and targets set by the owner. Targets developed by the building owners are displayed for comparative analysis against actual performance metrics in a process also known as monitoring-based commissioning.

Examples of poorly implemented dashboards abound. Recently, we visited a local university that completed construction of a new branch campus with aggressive building performance goals. As part of the project, they installed a large photovoltaic solar array. They also installed an energy dashboard with a large touch-screen monitor for the new buildings. We walked up to the dashboard with the program representative, and he proudly showed our group the new toy with all the bells and whistles. When the icon for electricity was selected, a graph appeared showing everyone the current consumption of electricity per building in kilowatt-hours. Similar icons were available for gas and potable water. Also, a separate icon displayed the generation of electricity from the PV array, with graphical displays. Next, we selected icons for temperature and relative humidity. When selected, those icons displayed graphs for temperature or humidity for the day, week, or month, as desired. We were also able to toggle between wet-bulb and dry-bulb temperatures. You can imagine the display.

When we asked the dean if the building's operational performance was aligned with the goals and targets set during design, he said he didn't know. When we asked if they knew the current site EUI of the buildings, he could not answer the question. We didn't want to seem critical of the new energy dashboard, so we decided to change the subject for the final question. We asked about access to the data displayed on the dashboard. The dean stated that they did not have easy access to the data. In fact, no one at the school and no one from the faculty or student cohort was able to freely access the data. The only way to get data from the system would be to submit a form to the facilities management department; a few days later someone from the department would respond to the request with an e-mail containing a data file with the exported data. The new building had the latest in building management systems, energy and water consumption meters, indoor air-quality sensors, outdoor climate and weather stations, and other meters and sensors, yet no one had any idea whether or not the building was performing as designed.

Our simple inquiries seemed to resonate with the dean, as if this was not the first time someone had asked questions about building performance. The dean seemed somewhat frustrated with the system, and we gathered from his answer that this data issue was a real problem for their group. As a university project, the primary objective of the new campus was to use the building to teach students. However, as a learning laboratory, there was clearly still work to be done. This is a common example of a system coming up short of the full capabilities represented by the technology of a dynamic whole-building dashboard.

A well-planned IoT-based integrated sustainability dashboard does more than simply visualize real-time trended data. It does more than show real-time trended data against industry-standard static thresholds or historically trended data. This kind of evidence-based performance enables you to know whether you are winning or losing against your key performance indicators in real time. The technology exists today to solve this problem. In fact, most projects already use it to make decisions during planning. Whole-building energy modeling is an industry-proven tool and is reliable if used properly. Consider that most project teams currently use energy models to select and size equipment or make decisions in design; they write a report for the building owners and then the model is never to be seen again. This transactional approach to energy modeling requires a significant investment of time and money that begs to be repurposed. Capitalizing upon the investment in the whole-building energy-modeling process and continuing to update the model during the later design phases and during construction is a logical extension that has the potential to change the way building owners invest in their buildings over 50, 75, and 100 years.

For example, creating the transactional model requires the operator to create the virtual environment, including the fabric of the envelope, fenestration, mechanical systems, lighting, and occupant profiles and schedules. Continuing to align the model during construction for submittals, change orders, value engineering, and quality-assurance field-testing is extremely beneficial to maintaining project goals during construction and later during operations. Post-construction, using the existing performance from utility meters and indoor air-quality sensors to pursue calibration is beneficial to closing performance gaps.

Repurposing the whole-building performance model and fine-tuning the model during construction is necessary for connecting design and construction to operations, for the purpose of ultimately delivering building performance (see, for example, fig. 7-8). The integration of smart building infrastructure with simulation allows owners to find and close gaps in performance while also going through commissioning of the building to confirm performance levels.

All too frequently, decisions are made during construction that have significant impacts on building performance. Without the common thread of a building performance model, owners and project stakeholders are left with only intuition and guesswork. Further, precise construction of building details is mission-critical for high-performance projects. Many projects are moving beyond typical construction administration work by facilitating complete envelope commissioning and comprehensive quality control testing. The results of field-testing must be weighed to determine and mitigate the impacts on building performance.

Commissioning and field inspections are a technical verification process and quality check for buildings systems. They provide a baseline for performance and ensure that design and building targets are met. Ideally, a third-party firm is contracted as your agent to commission the building, and commissioning agents are retained early in the design stages to document design intent and review drawings, and then they continue their work through the first year of occupancy to verify performance.

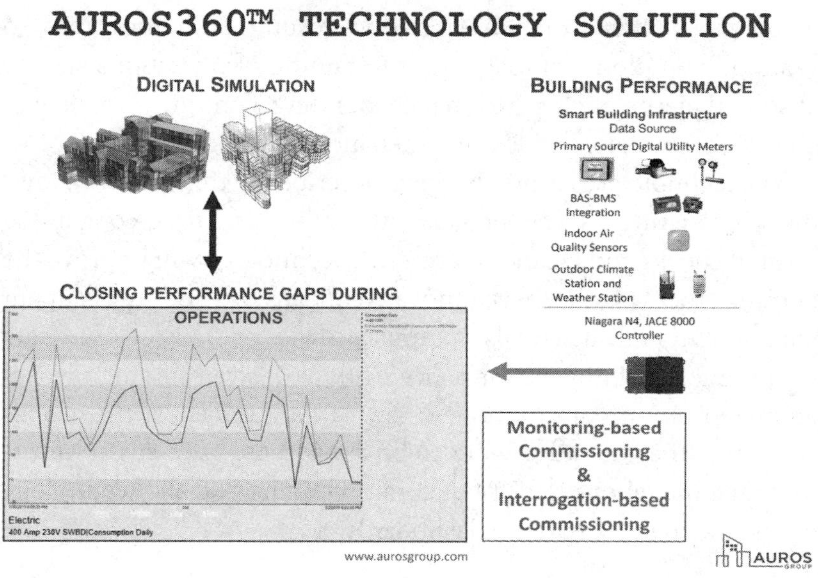

Figure 7-8. Integration of Evidence-Based Performance.

Comprehensive quality-control testing of envelope enhancements includes air-seal training and observation, pre-insulation and final blower-door testing and air-seal troubleshooting, and post-insulation and final thermal image testing to ensure that the high-performance design is implemented correctly. Additionally, duct-leakage tests and final test and balance of the HVAC system are critically significant in any deeply integrated energy-efficient project. This quality assurance guarantees durability, performance, and construction quality, allowing project engineers to design close to their calculations and optimize the size and capacity of their proposed systems.

Best practices during construction require commissioning agents to monitor, verify, and fine-tune the building's systems for optimal performance. Further, commissioning agents should provide building operators with an "owner's maintenance manual" to keep buildings running in optimum condition. Best practices recommend that building operators conduct recommissioning after two years of operation, complete with retraining of operations staff to ensure that the high-performance building continues to deliver efficiency, comfort, and low maintenance, as designed.

The technology platform we have described in this chapter allows owners to wring out the inefficiency of costly and time-consuming form of checks and balances. The same technology will improve the current process with a new, innovative means of building occupant engagement, and it will also permit knowledge transfer, eliminate lag time for building performance drift, and save building owners an incredible amount of money. Few can argue against the proposition that there is much work to be done in the built environment to merge the labor-intensity of commissioning and the technology of integrated sustainability dashboards.

Benefits of Smart Systems

The benefits of a whole-building energy model used during operations cannot be overstated. Interrogation-based commissioning and monitoring-based commissioning capabilities have the potential to capture multiple benefits of energy efficiency and to scale the practice of designing and delivering high-performance buildings.[6] As we are introducing new terminology to the field, we define "interrogation-based commissioning" as comparing digital simulation to trended performance for the purpose of clarifying and narrowing gaps in performance for all forms of energy and indoor environmental quality. Performance dashboards capable of interrogation-based commissioning are ones that integrate predicted building performance metrics with trended building performance.

Interrogation-based commissioning services using sustainability dashboards are a cost-effective extension of the technology to identify the causes of performance gaps in energy consumption and indoor air quality. Traditionally, building owners use labor-intensive retro-commissioning services to identify and close performance gaps. Once the integrated sustainability dashboard is implemented, building owners can oversee and manage dozens of integration-based simulations to identify likely root causes of gaps in performance. The results of each simulation can be easily viewed on the integrated sustainability dashboard against performance trends.

Considering that a good energy modeler can run dozens of simulated scenarios over the period of a week, building owners now have access to a tool that can quickly identify root causes of gaps in performance. This can save a lot of money that would otherwise be spent on field investigations to identify performance gaps; traditional commissioning services can range from upwards of tens of thousands to hundreds of thousands of dollars. Clearly, using technology to identify root causes is far more cost-effective.

Monitoring-based commissioning services allow decision makers to import, manage, and analyze building trended data and operational whole-building model simulations in one digital platform. Actual consumption data is compared to the simulation model to enhance building performance. One of two outcomes will result: either buildings will close their performance gaps, or the operational whole-building model simulation will be improved for future years of the building life. Both outcomes work to the benefit of the building owner and project stakeholders. Future monitoring-based commissioning operations can be used to

1. Undertake building occupant engagement and post-occupancy evaluations.
2. Aid in delivering a seamless handoff from construction to building operation.
3. Investigate the impact of future building improvements using real building data.
4. Improve operational models for future performance contracting.
5. Correlate digital meter and sensor data.
6. Comply with sustainability certification program criteria.

Owners, public and private funding resources, and stakeholders who decide to invest in building performance do so because they know the value of the asset and the value to their employees and the people in the buildings. IoT-based smart building infrastructure and integrated sustainability dashboards need to reinforce investment decisions by providing proven feedback loops that are aligned with investments in building performance. We propose that the return on integration (ROInt) for a project consider financial, environmental, and human health and productivity as returns on the investment and payback periods.[7] An example of this is found in the combined costs of a dashboard, modeling, utility meters, IAQ sensors, and a weather station for a project, altogether coming to ~$143K—a small portion of an overall $3M budget for a project. When we looked for impacts on the annual energy expenses for this ten-story building, we learned it costs $306K for 2.549 MWh of energy. Our approach to the power of existing buildings enables a realistic 70 percent reduction in annual energy consumption while setting the owner on a path toward Zero-Energy. The retrofit costs of $3M, focusing on the envelope and windows, also includes the dashboard. The single bottom-line cost-savings have a payback of 14 years. When we also add the social cost of carbon (SCC) at $40/ ton of environmental and social impacts avoided (tonnes of CO_2e derived from the number of kWh avoided, multiplied by the CO_2 emission per kWh), the payback period is reduced by two years. When only looking at energy without including SSC, retrofitting doesn't get you all the way to Zero-Energy. With a deeper dive into measurement and occupants, including impacts on human health and productivity, the resulting feedback loops of data only shorten payback periods and increase ROInt.

We know this because high-performance buildings increase productivity and reduce absenteeism. While these costs are relative to any building and its occupants, they are more tangible in businesses and a bit harder to get at for buildings such as schools. An important thing to keep in mind is that employers spend 92 percent of their annual operating costs on people.[8] Thus, investments in indoor environmental quality such as WELL building standards result in meaningful returns on investment, including attracting and retaining employees, building brand equity, improving performance, promoting health and well-being, and eliminating wasted time and resources. Absenteeism is more costly than you may think. According to "Absenteeism: The Bottom-Line Killer," a publication of the workforce solution company Circadian, unscheduled absenteeism costs roughly $3,600 per year for each hourly worker and $2,650 each year for salaried employees.[9] This adds up quickly as you multiply the number of these types of workers in a building by these costs, and then add this to your ROInt. Absences cost UK businesses an estimated £29 billion annually, and the average number of days lost is 6.3.[10] You can take this even further if you're looking for the benefits of going green.[11] A national review of thirty green schools demonstrates that green schools cost less than 2 percent more than conventional schools—or about $3 per square foot ($3/sf)—but provide financial benefits that are 20 times as large. The financial savings are about $70 per sf, 20 times as high as the cost of going green. Research at Carnegie Mellon University has shown that single bottom-line returns (only looking at financial capital) on investment, which normally take multiple years, are achieved in months when factoring in environmental impacts avoided and improvements in human health and productivity.[12] If you include these multiple benefits in your plans for renovating an existing building, what's your ROInt?

Conclusion

As technology allows us all to become smarter and more connected, your building should also be part of this wave of innovation. It is now possible to visualize building performance in dashboards, and we can know that for every dollar invested in performance-measurement technology, there are financial, environmental, and human returns on integration (ROInt). A properly planned, IoT-based integrated sustainability dashboard enables real-time monitoring, which facilitates reducing energy consumption, improving indoor environmental conditions, and monitoring occupant behaviors. The integrated sustainability dashboard should mirror the key performance indicators based on the goals set in early planning. You now have a road map to help your team develop a smart building system infrastructure. The steps necessary for this infrastructure are summarized in table 7-1.

Before turning the page and going on to chapter 8, look at the homework opportunities below and apply learning from this chapter to your own building project. When you get a few moments, close your eyes and consider yourself and your team successfully completing your building renovations and projects. Now visualize what you would want your existing building's dashboard to look like. What do you see?

Project Development Homework

- ✔ What are the building systems (meters and sensors) that can provide feedback for critical key performance indicators necessary to demonstrate the conversion of your existing building to a high-performance building?
- ✔ What IAQ sensors will enable you to prove success to stakeholders? Are these sensors reliable and can they be easily recalibrated to ensure continuous data reliability?
- ✔ Write up a brief description of *how* repurposing performance and building meters and sensors will build the business case to extract integrated value in your existing-building project. Use this information to tell a compelling story about *why* investments in your building are necessary.
- ✔ Develop a mental model of your future dashboard and sketch it on paper or on a tablet.
- ✔ Identify the data necessary to operate high-performance buildings.

Table 7-1. Smart Building System Components and Stepwise Approach

Steps	Equipment Advocate Services	Connectivity Enabled
Where to Start— Baseline	Utility Meters	Baseline current performance including historical utility data
Planning	Modeling	Collaboration, stakeholders, integration of dynamic data
Collaboration	IAQ Sensors	Continuous monitoring, daily averaging, comparison to targets and standards
We Can Afford This	Weather and Climate Stations	Weather forecast to correspond with owner-based goals and targets
Envelope	Modeling	Monitor, verify, and fine-tune; commissioning, comprehensive quality-control testing
Net Zero	Modeling	Whole-building energy modeling, scenario assessment
Operating for High Impact	Dashboard	Operationalized whole-building energy model used during operations; occupant engagement

Chapter 8:

Case Studies

Progress is impossible without change, and those who cannot change their minds cannot change anything.
— George Bernard Shaw

I N THIS CHAPTER WE LOOK AT A few case study projects and the context for building the business case for each, showing decision makers of existing-building projects that they *can* afford this. We chose case studies that show a few different building types. We had a direct hand in these projects as advocates for the transformation of the buildings, the development of the projects, and the collaboration of teams. We were able to work with these building owners and stakeholders to find the business case and meaningful impacts of these existing-building projects.

Three Rivers Luxury Residential—Pittsburgh, Pennsylvania
Three Rivers Towers is an office and residential building completed in 1964 in downtown Pittsburgh, Pennsylvania, next to mass transit systems in a metropolitan area with a significant

amount of building stock around fifty years of age or older. A twenty-six-story condominium complex, Three Rivers has decades of deferred maintenance and is in need of deep energy renovations. The complex is over 50 percent occupied by residents and owners, so executing a holistic approach to building restoration requires an almost perfect plan.

Three Rivers considered various approaches to the restoration of an owner-occupied residential building—like Passive House's EnerPHit strategy. EnerPHit is a program for certifying energy retrofits with Passive House components for existing buildings to realize energy savings between 75 and 90 percent using improved thermal insulation, reduced thermal bridges, improved airtightness, high-quality windows, ventilation with heat recovery, efficient heating and cooling generation, and use of renewable energy sources. EnerPHit is a strategic pathway to deeply renovate any existing building for low energy use and high indoor air quality at little to no first-cost premium. EnerPHit is unique in that it is effective whether an owner chooses to restore a vacant building from the inside out or restore an occupied building from the outside in. In the case of an occupied restoration, the building envelope is reconstructed from the outside of the building, as it wraps the building, to create a high-performance envelope. This technique is particularly useful when it is impractical to temporarily relocate residents.

DOES IT HAVE POTENTIAL?

Three Rivers Towers is a steel-framed building clad with precast concrete panels and metal panels into which windows are set. The project concept at the Towers started with a need to address its chilled-water source from a neighboring third-party building. The costs for chilled water are exorbitant and the infrastructure is antiquated. By all accounts, costs for chilled water are going to continue to increase in the near future. We also discovered significant deferred maintenance and life-cycle issues at the building. The R-value of the envelope was less than R-10, and there was no air barrier. The R-value is an insulating material's resistance to conductive heat flow. It is measured in terms of its thermal resistance, or R-value. The higher the R-value, the greater the insulating effectiveness. The R-value depends on the type of insulation, its thickness, and its density. Beyond the lack of insulation, the life expectancies of the exterior windows, mechanical systems, façade, roof, and other systems had expired, with many systems failing completely. Ownership, understanding that something needed to be done, had begun to investigate opportunities, weighing incremental restoration versus a holistic approach using building science and goal-setting. This opened up a conversation about value. A holistic approach takes an old dysfunctional building ("zombie high-rise") and solves a majority of the life-cycle problems while simultaneously reducing energy consumption up to 90 percent.

WHERE TO START

The logical place to start the project was to hire feasibility consultants to identify building-wide performance baselines. The baseline utility expenses provided an understanding of annual usage and maintenance costs. The baseline measure of indoor environmental quality, or lack thereof, showed a vast opportunity for the improvement of air, light, and sound quality. The business case for an incremental approach could be weighed against a holistic approach over longer timelines that could stretch out for fifteen years. The risk of catastrophic failure could be taken into account as a catalyst for change and risk management.

PLANNING

The feasibility study naturally led to baseline assessments, and then to a goals-based strategy to help solve the problem(s). We then refined the long-term strategy to address annual utility costs and understand what needs to be done to meet goals. In this case, we proposed working with a proven envelope solution, Passive House EnerPHit, and a team of experts in the field.

The internal project team included the condominium board and facilities-management team. The external team was composed of two of the authors of this book, a local architect, and an expert in the Passive House standard. The holistic approach articulated by the team made clear the superior business case of a long-term, goals-based plan. The project team included experienced architects specializing in Passive House design and implementation, along with performance advocates to set goals with the owner for the development of a building performance model to guide building improvement choices. Multiple elements of this project remain in the conceptual phase as this project develops. To convince a group of owners to invest in holistic building restoration requires data, a well-thought out plan, and a short-, medium-, and long-term budget that ties together the physical needs of the building with a financial realities of a condominium association.

Can They Afford It?

Business as usual, using an incremental approach, would be roughly $30M to $40M in incremental costs over the next ten years. Annual operating expenses of $850K would remain unchanged. A holistic PHi retrofit approach would total $20M to $30M invested over two to five years, and the annual operating expenses of $850K would be reduced by 80 percent. Multiply this by fifty years, and it becomes clear that taking action on this holistic approach now would save $34M and help tunnel through traditional diminishing returns to find larger whole-building system benefits.

The Envelope Holds the Key

By addressing the envelope first, the Towers owners are able to reduce building mechanical loads, resulting in smaller and decoupled systems for ventilation, heating, and cooling.

Improvement in the thermal barrier and elimination of thermal bridging directly impact thermal comfort for building occupants. Also, reducing exfiltration and infiltration of outside air while simultaneously increasing filtration ensures improved indoor air quality. As an EnerPHit project, the Towers may reduce consumption of energy dramatically from a current site EUI of 70 kBtu/sf/yr to a simulated site EUI of 20 kBtu/sf/yr. At this level, offsetting the remaining building use of energy using renewables, on-site generation, and other means is much more feasible.

OPERATE THE BUILDING FOR MAXIMUM BENEFIT

Examples of other successful EnerPHit projects include the Urban Green Council in New York City; a 140-year-old Victorian stone building in Gloucestershire, UK, converted into a youth hostel; a multistory residential tower in Portsmouth that remained occupied during renovation; and a number of other projects in Linz, Austria, the United Kingdom, and Sweden. There was a proposal to Penn State University about using this approach on an affordable-housing building. Penn State has a cooperative agreement with the United Nations Global Building Network to provide research and development for Sustainable Building Goals (SBGs); they are looking for ways to support the UN's Sustainable Development Goals (SDGs), and this project would touch at least five of the seventeen SDGs: Good Health and Well-Being, Affordable and Clean Energy, Reduced Inequalities, Sustainable Cities and Communities, and Climate Action. This approach and our capabilities as performance advocates show the potential for reductions in energy expenses with dramatic improvements in indoor air quality in a community's worst-performing building, where some of that community's least-affluent citizens live. Add to that the interest of developers, energy companies, and utilities to help finance these types of projects, and existing-building projects often align with the interests of a large base of stakeholders.

Environmental Charter School (ECS)—Garfield, Pennsylvania

The Environmental Charter School had a need to find and develop a new middle school for grades 6–8 and a 9th-grade academy in Garfield, near Pittsburgh, Pennsylvania. This building required a significant investment and full renovation of a vacant school building that had been unoccupied for years. Project goals include meeting the Architecture 2030 Challenge, WELL Building Silver certification, and RESET Air certification using Passive House principles and strategies.

Because ECS is a K–8 charter school focused on building systems from the perspective of the environment and ecology, they had a very specific interest in using their buildings as part of the curriculum. The team on this project felt a strong need to understand how best to link the building to the ECS mission of "Growing Citizens." We began with a goal-setting charrette and immediately followed that with the creation of a whole-building performance model and a detailed estimate comparing code-based plans with new conceptual plans. The purpose was to develop a baseline of benchmarks for energy use, indoor air quality, first costs, and long-term operating costs. We represented the owner as their performance advocate and provided the team with all the information they needed to make the highest-value decisions throughout the project, delivering the highest-quality building at the lowest-possible costs. We specified open-integrated measurement and verification systems to provide real-time automated feedback of building performance based on the key performance indicators defined by the owners. Owners will know on day one of operations whether or not they achieved the performance goals they set in early planning.

DOES IT HAVE POTENTIAL?

The aspirational targets for the building are mission-aligned with the environmental school's pedagogy and sustainability initiatives. The school plans to incorporate the building into their STEM curriculum as a learning environment. They believe in utilizing the embodied energy in an existing building and were excited about rehabilitating an existing facility as opposed to constructing a new one. The school was pleased to have empirical data supporting their desire to reach high performance without any increase in construction costs.

WHERE TO START

The school began thinking about its operations using WELL Building Certification (see chapter 1, box 1-1). WELL Building has proven to be compatible with the school's focus on building occupants. Given the amount of time we spend inside, we know that buildings have a direct bearing on occupant health and performance. WELL Building is an evidence-based performance standard measuring, certifying, and monitoring building features that influence human health and wellness within seven concept categories (i.e., air, comfort, fitness, light, mind, nourishment, and water). These performance metrics align with Science, Technology, Engineering, and Math (STEM) and can be used in curriculum. For school administrators, investing in an environment conducive to the education of children is an easy decision. However, they also wanted to measure and monitor the impacts of improved indoor environmental quality for student performance and health. So, they added RESET Air certification and an integrated building-performance dashboard to their list of goals.

PLANNING

As advocates for this building and the project team, AUROS Group's services included owner's representation, performance advocacy, sustainability certification program support, cost-estimating, budget control, operational whole-building modeling, smart building infrastructure and support, the development of building commissioning specifications and detailed specifications for measurement and verification systems. The early use of a design charrette and discovery planning processes were vital to developing a strategy to validate indoor environmental quality during operations.

CAN THEY AFFORD IT?

It's one thing to develop aspirational benchmarks for energy use, indoor air quality, first costs, and long-term operating costs, but quite another to pay for it. The school had a hard finance budget of $11M, with little flexibility to increase the budget. The developer provided the upfront capital as part of a twenty-year lease-to-own agreement for the school. This project resulted in the school reducing its utility consumption by half. This is significant, as the lease agreement requires them to pay for their own utilities. The project is projected to save operational costs of $432K in energy costs over twenty years. The social cost of carbon (SSC) savings for carbon offset add another $381K in savings based on social and environmental impacts avoided. The school was fortunate to find a developer who was visionary and willing to provide financing for this kind of innovative project.

THE ENVELOPE HOLDS THE KEY

The school tested numerous envelope assemblies, but ultimately balanced and optimized its envelope against planned mechanical systems. The mechanical systems were selected to comply with WELL Building requirements, which meant solving rigorous ventilation needs. Understanding their needs in order to meet their metrics for ventilation, heating, and cooling helped them to land on a strategy for systems. At that time, the school tested numerous thermal barrier values to balance system sizes, ending up with an optimum value for their thermal barrier and final system sizes. This strategy ended up saving the owner about $400K–$500K in construction costs by avoiding the mechanical system costs of oversized systems. Additional benefits included reduced infiltration and exfiltration of outside air while simultaneously increasing filtration, ensuring improved indoor air quality.

The project baseline information established the building's site EUI of 72.4 kBtu/sf/yr. The improved design resulted in a site EUI of 47.7 kBtu/sf/yr. The annual energy costs were reduced from $76,565 to $56,495. At this level, introducing renewables, on-site generation, and other means in the future to offset energy consumption is feasible.

OPERATE THE BUILDING FOR MAXIMUM BENEFIT

This project will have an on-site integrated sustainability dashboard that will engage occupants by showing energy savings and performance across five indoor air-quality parameters.

The places and spaces where children learn and play are important. Several powerful teaching opportunities can be incorporated into this project: learning how to transform a vacant building into a high-performance school, understanding that lower energy usage and superior indoor environmental conditions are not mutually exclusive, and understanding that high-performance building renovation does not necessarily cost a premium.

Transitioning from setting goals and targets to testing them out in enhanced modeling and simulation environments demonstrates what is possible. This process enables stakeholders to thoughtfully connect the building's operations with the school's curriculum. The school has already integrated social equity into their environmental curriculum, yet they had not tapped the potential of their own built environment for teaching lessons about water, energy, and air quality. The transformation of an old building into a high-performance school was a key component of a transformational learning experience for all stakeholders, not just students. We are proud to have facilitated the alignment of the owner's stakeholder team by providing the information they need to make the highest-value decisions throughout the project, resulting in the highest-quality building at the lowest-possible costs.

Rockwell Hall, Duquesne University, Pittsburgh, Pennsylvania

Rockwell Hall is home to the university's top-ranked MBA Sustainable Business Practices program and business school. The building is over fifty years old and encompasses 165,000 square feet. Its size and age make it representative of over 5 million existing buildings in the United States. The MBA Sustainability program uses its existing building as a living laboratory in its curriculum to test a working master plan, find innovative solutions, and support renovations and improvements with the business case linked to the UN SDGs.

The decision makers included two deans of the school and multiple cohorts of MBA students, faculty, and facilities-management people. What started as a LEED Commercial Interiors design competition in 2009 has led the decision makers to reimagine the building as a learning laboratory that connects utilities, systems, and sustainability with the objective of showing that for every dollar invested in this existing building, we can show a financial, environmental, and social return on integration (ROInt).[1] The competition runs on an annual basis and has morphed into an energy-management-systems design competition.

DOES THE BUILDING HAVE POTENTIAL?

In addition to functioning as a learning laboratory for the business students, faculty, and facilities staff, this building has the potential to be the proof of concept for the other forty-nine buildings on Duquesne University's campus. We saw the potential for whole-building modeling to test renovation plans, provide financial analysis, and provoke new insight into projects that are within the control and understanding of decision makers.

WHERE TO START

The building was already scheduled for single-floor renovations. Why not educate decision makers as to the potential environmental and social impacts of existing-building renovations? It seems appropriate that one of the top-ranked sustainability MBA programs in the United States is located in a high-performance building functioning as a learning laboratory. One of the authors of this book teaches there, which provides a logical starting point.

PLANNING

We developed a sustainability plan using whole-building modeling and integrated utility meters, indoor air-quality sensors, and a dashboard to track and compare building improvements with other similar projects on campus, while MBA students participated in an annual energy-management-system (EMS) and renovation project.

As a faculty member actively engaged in sustainability business practices, research, and an alum with decades of construction and whole-building design experience, the authors routinely engage students, administrators, and facilities project managers to find unique solutions. Program faculty, students, and alumni have all developed methods that demonstrate cost-savings and cost-avoidance and have aided the development of a strategic masterplan, building out smart building infrastructure with the cooperation of facilities management and other on-campus stakeholders to work toward common goals through university based research projects.

CAN THEY AFFORD IT?

Prior student cohorts, along with industry partners, had already demonstrated numerous energy-conservation measures and bundles of measures that could transform Rockwell Hall. Working with other industry partners, a current cohort developed detailed estimates for the measures. This foundational information was critical to unlocking the MBA sustainability program's ability to create financial models that support and validate affordability. For example, we know that a full Passive House EnerPHit strategy at Rockwell Hall will cost between $3M and $3.7M, depending upon a few variables. Teams of MBA students calculate numerous financial models that incorporate an integrated bottom line, including environmental and social impacts along with financial performance. Simple financial-performance models are developed using various return-on-investment calculations that incorporate social cost of carbon (SCC) and demonstrate bigger benefits from tunneling through traditional cost barriers. Graduate students in the MBA sustainability program take part in an annual design competition for renovation solutions, with returns determined by simulation and whole-building modeling in which they can show that for every dollar invested in the renovation, there is an environmental and social return on the investment.

THE ENVELOPE HOLDS THE KEY

Rockwell Hall is a steel-framed building clad with precast panels into which windows are set. We discovered significant deferred-maintenance and life-cycle issues in the building. The R-value is less than an R-10, and there is no air barrier. Additionally, the life expectancy of the exterior windows, mechanical systems, façade, roof, and other systems had expired. Facilities management, understanding that something needed to be done, had a building plan that included incremental improvement to assemblies and systems. Unfortunately, many of these plans included in-kind replacement strategies. For example, as floors are renovated, the exterior walls are either left in place or replaced in-kind. As hard as it is to believe, facilities management will remove an exterior wall (R-8 with no air barrier) and replace it with an R-8 and still with no air barrier. This lost opportunity to leverage the costs of replacement is illogical, but many building owners make similar decisions when the optimum envelope condition is not designed. The MBA sustainability program is making progress in identifying the benefits and affordability of a holistic approach backed by building science and goal-setting. This opens up a conversation about value.

Zero-Energy is a challenge in the short term, given the state of prior renovations and lack of an integrated design in the past, but goals for very low site EUI can become part of the building's plan. The building was pressure-tested and utilities were taken into account, which revealed a baseline site EUI of 157 kBtu/sf/yr. An EnerPHit strategy has a goal of reducing this to a site EUI of 20 kBtu/sf/yr.

OPERATE THE BUILDING FOR MAXIMUM BENEFIT

With a robust and durable smart building infrastructure of meters, sensors, and IoT-connected technology platform, the Rockwell Hall project allows for testing of ideas and options for energy efficiency and indoor environmental quality improvements. The cohort continues its planning for cost-reductions and improvements that can be rolled out to other buildings on campus.

As a living laboratory, Rockwell Hall is used in curriculum and pedagogy to introduce future business leaders to whole-building management, capital expenditures, and decision-making. With over 100,000 MBAs graduating each year most will never be challenged to see how the buildings they live and work in can save money, reduce impacts on the environment, and improve the health and productivity of occupants. There are over 3,000 colleges and universities in the United States, all with the potential to replicate these efforts while finding their own benefits. The ideas from the graduate student design competition can be mapped onto UN SDGs, creating a vision for how changes in this building can be applied to benefit stakeholders anywhere in the world.

For Rockwell Hall, we know that the Passive House EnerPHit approach to thermal barriers, including windows and airtightness, will cost $3M–$3.7M, with a payback period inside of three years. We are still metering all utilities and indoor air quality and have good assumptions for a baseline assessment. Most meaningful, given the use of the MBA student EMS design competition, is that renovating existing buildings does not have to be an exact science. The simple fact is that the numbers for value are large, as compared to the costs to transform and renovate old buildings into high-performance buildings.

Most people are overwhelmed by poor building performance, assuming that the expenses and difficulties to remedy it and transform their building are simply unaffordable. Instead, we recommend that the building owners take a step back to see the strategic opportunity for value maximization as they are seeking and developing innovative building solutions. We have had sociologists take students through building spaces to see what impression the space provides as they come into it. Is the space inspiring, bright, and interactive, or is it dark, cold, and closed off? For students and faculty who spend many months and years inside a building, the more we can integrate their experience with the built environment into curriculum and into STEM programs, the more valuable buildings are as tools for experiential learning. Then, with this learning, our renovations of existing buildings help to save the world, one building at a time.

Conclusion

Here in chapter 8, we've taken a step back from the format of the preceding chapters to review three evidence-based case study projects. In doing so, we have provided a context for building a business case for each case study while elevating the potential of these projects beyond a myopic focus on lowest-cost renovations to include a more dynamic systems-thinking and whole-building-solution approach to existing buildings. The goal in reviewing these very different case studies is to show the purposeful approaches you can take to realizing the potential of your own building. We also want to show decision makers that you *can* afford this and that there are strategic opportunities for buildings beyond first costs. While there are more projects we could have looked at, these three cases and a few other covered in the book represent a variety of building types and uses where we were advocates for the transformation of these buildings, the development of the projects, and collaboration of teams. Keys to success in any existing-building project are to see the importance of representing the owner's and occupants' strategic, long-term interests in the building, advocacy for improved performance including human health and productivity, seeking sustainability certification programs, operational whole-building modeling, and smart building infrastructure. These integrated elements of existing-building projects, along with a vision of a sustainable, healthy and productive society, will ensure that you find the full power of your own building while making informed whole-building decisions.

Project Development Homework

✔ Review these cases in light of your own building(s). What strategies and ideas can you reuse in your planned project? Where do you start? How will you go about planning? Can you afford this? What will you do about the envelope? How will you operate the building for high performance? Remember that each step offers a real opportunity to build upon the innovative ideas of others to transform your building.

Chapter 9:

Existing Buildings Can Save the World

It's only when you hitch your wagon to something larger than yourself that you realize your true potential and discover the role you'll play in writing the next great chapter in America's story.

— Barack Obama

Tᴴɪꜱ ʙᴏᴏᴋ ᴡᴀꜱ ᴅᴇꜱɪɢɴᴇᴅ ᴛᴏ accomplish one goal: prove to building owners of existing buildings that it is possible not only to make old buildings perform like new ones but also to do that without paying much, if any, premium in construction costs. The primary benefits of accomplishing that goal are obvious: a dramatic reduction of energy consumption and a dramatic improvement of indoor environmental quality. But we haven't yet told you the rest of the story. We haven't shared with you how the choices that building owners make every day have lasting consequences and contribute to either extending the life of this planet or to destroying life on this planet. At this point in world history, every building owner has an opportunity to put their fingerprint on the future of our world.

We couldn't have started the book with this thesis, or we would have lost hordes of rational, bottom-line-oriented building owners and project teams who would have concluded that we are just some wild-eyed sustainability zealots. We first had to lay out the business case, the prevailing building science, and recent technological breakthroughs that substantiate owners' decisions to invest in making old buildings perform like new ones. Most owners and developers don't have capital to invest in fantasies. As we say throughout the book, investing in the efficiency and the performance of an old building might appear fanciful if it didn't make so much business and financial sense. But there is so much more to consider, because each owner's discrete investment decision, when aggregated with other such decisions, results in widespread global consequences. Before we begin that discussion, let's first summarize the business case, prevailing building science, and modern technologies that make holistically investing in old buildings almost irrational to ignore. This is the path to making the case for your building and realizing its full potential (see fig. 9-1).

Step One — Recognize the Potential of Your Building

Do not look at your building as a money pit. Building owners today are rarely given post-construction data to validate whether or not they got what they paid for in the renovation of their buildings. While there are building performance assumptions built into every pro forma, that is, the methods for calculating financial results for certain projects, often used to justify investment in new or existing buildings, rarely, if ever, are owners/developers given proof that they have achieved the performance in which they invested. From a building science perspective, existing buildings have the same opportunities as new buildings to reach ultra-low energy consumption and ultra-high indoor environmental quality. We recommend starting by identifying baseline performance measurements for current energy consumption and indoor environmental quality. With baseline performances in hand, the team

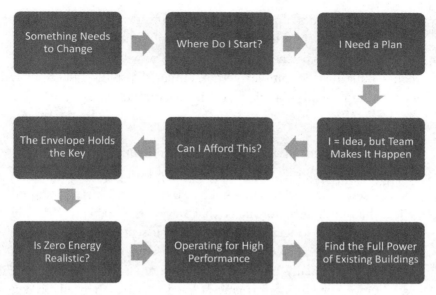

Figure 9-1. Pathway to the Power of Existing Buildings

can begin evaluating an existing building's theoretical optimum level of performance in terms of energy consumption and indoor environmental quality.

Step Two — Know Where and How to Start

When considering investments in existing buildings, most owners and developers wait until they have a crisis or even complete failure in their systems, at which point they usually hire an engineer and then maybe an architect to replace their systems in kind with little to no improvement in building performance. We must instead consider life-cycle, deferred maintenance, and planned renovation triggers as opportunities to transform building performance. In many circles, this kind of thinking is becoming a best practice. These trigger opportunities help owners and project teams evaluate a wider array of building improvement investments using a holistic analysis of costs versus benefits. We make the case that existing buildings have hidden sources of value that, unless considered holistically, will never be realized. In addition, occupants of buildings today expect those buildings to be energy efficient and to provide healthy environments. Buildings are complex with interrelated systems, but each has its own unique path to performing like a new building.

Step Three — Understand the Importance of Planning

A holistic approach to renovating existing buildings requires that owners set clear performance-metric targets; the metrics should reflect the building performance that an owner is expecting during operations. It's no longer enough to merely *design* a high-performing renovation; building owners and occupants today expect proof that a building is actually performing in energy-efficient and healthy ways. Once specific performance targets are set, then all project team members must be aligned and fully committed to delivering those metrics in operations. The technology exists today for an owner to know, post-occupancy, that their building is performing to the projected financial pro forma expectations. Teams need to see how tools and technology will be used to connect design to operations, ensuring at every step in the construction process that the performance goals are reinforced and supported.

Step Four — Collaborate to Bring About Change and Innovation

An organization's culture, embodied in the attitudes, opinions, and beliefs of the owner, determines the likelihood of success for an aspirational sustainability agenda for any project. The owner must be convinced that the project team members each share his or her values in terms of both sustainability and the accountability of each member to deliver against the owner's goals. This is the most important social component of any deep building renovation and will either propel a project forward to success or ensure its failure. There is no substitute for experience. Your key team members must be able to show examples of how comfortable they are with the transparency and collaboration required to achieve aggressive performance goals. Owners are looking for team members who talk more about actual performance in operations than they talk about sustainability standards and certificates. Owners should evaluate every team member's body of work, paying specific attention to verifiable examples of the achievement of performance-based goals.

Step Five — Determine If You Can Afford to Restore an Existing Building

Effectively articulating the business case for a project requires one to merge the financial and social needs of an organization. Many business cases may meet the threshold of an organization's investment hurdles, but the ones that get funded provide significant confidence in the accuracy of the predicted costs, risks, and benefits of the project. In terms of both costs and risks, it's important to document cost reductions, cost increases (if any), and cost avoidances. Amory Lovins provides this same kind of insight when reminding us that "by skimping on design, the owner gets costlier equipment, higher energy costs, and a less competitive and comfortable building; the tenants get lower productivity and higher rent and operating costs."[1]

Step Six — Focus on the Envelope

A holistic approach to making old buildings perform like new buildings in a way that is affordable is being unlocked and enabled today by the rapid adoption of Passive House building science around the world. Applying the Natural Order of Sustainability—passive first, active second, renewables last— provides a building-science-based methodology to reach the highest-performing building at the lowest possible cost. Passive House is a systemic building science approach to commercial and residential construction. Focusing on the envelope increases insulation, airtightness, energy conserving lighting and equipment, and improved ventilation in order to hit aggressive performance targets for whole-building performance. Investing in the envelope pays and pays big if you look beyond just the reduction in energy usage to the further goal of improving indoor environmental quality.

Step Seven — Set Goals to Get to Zero-Energy

Ultra-low energy consumption and ultra-high indoor environmental quality are symbiotic goals for the enlightened owner or developer. It shocks many when we say that the goal of Zero-Energy is just as achievable for an existing building as it is for a new one. Our approach identifies how to get the most performance out of a building, given its constraints, and then determines the short- and long-term operational costs to owners so they can define what level of performance they want—that is, the level in which they would like to invest. Once the theoretical level of performance is targeted, the path to Zero-Energy will be the most affordable path possible. Usually with this level of data, an owner's confidence in the achievement of these goals increases considerably. Any owner trying to figure out how to reach Architecture 2030 Challenge goals needs a master plan that encompasses these first seven steps. Without a plan, most successful organizations will find that increasing energy loads from organic growth is outstripping their ability to incrementally reduce energy demand through sustainability measures and conservation.

Step Eight — Operate the Building for Maximum Benefit

"Did I Get What I Paid For?" This should be the clarion call for every owner and developer looking to invest in existing buildings. The proof of performance lies in continuous monitoring and dynamically displaying performance results for all types of energy consumption and indoor environmental quality. To do this work simply and cost-effectively, owners must consider their technology networking and information-management systems as early as possible in the design phase, while they are developing high-performance building envelopes. All holistic building performance work must be accomplished early in planning. It may mean investing a bit more in the early soft costs of a project. However, those investments will pay big dividends in less re-work, more aligned goals, an evidence-based process for managing change order requests throughout a project, and a clear view on day one of operations whether the building is achieving the goals as outlined in the project's financial pro forma.

At this point, you should feel a sense of relief and confidence that it's possible for old buildings to perform at the same level as new buildings. In fact, in many ways, old buildings have the "bones" perfectly tailored for goals like Passive House building science, RESET Air, and Zero-Energy. There is also evidence that cities' CO_2-reduction targets, over the coming decade, can be met simply by retrofitting and reusing existing buildings rather than demolishing them and then building new, efficient ones.[2] But you still need the rest of the story.

The Big Picture—Your Decisions Contribute to Powerful,
Global Consequences

In 2015, the United Nations adopted the 2030 Agenda for Sustainable Development and its 17 Sustainable Development Goals with 169 targets. With these new goals applied universally to all, over the next fifteen years countries can mobilize efforts to end all forms of poverty, fight inequalities, develop sustainable cities and communities, and tackle climate change, all while ensuring that no one is left behind.

GOAL 1: No Poverty
GOAL 2: Zero Hunger
GOAL 3: Good Health and Well-Being
GOAL 4: Quality Education

GOAL 5: Gender Equality
GOAL 6: Clean Water and Sanitation
GOAL 7: Affordable and Clean Energy
GOAL 8: Decent Work and Economic Growth
GOAL 9: Industry, Innovation, and Infrastructure
GOAL 10: Reduced Inequality
GOAL 11: Sustainable Cities and Communities
GOAL 12: Responsible Consumption and Production
GOAL 13: Climate Action
GOAL 14: Life below Water
GOAL 15: Life on Land
GOAL 16: Peace and Justice Strong Institutions
GOAL 17: Partnerships to Achieve the Goal

If you look at the UN's Sustainable Development Goals, you'll see that a number of them benefit from a commitment to investing in ultra-low energy consumption and ultra-high indoor environmental quality of existing buildings. The World Green Building Council (WBGC) confirms these positive impacts, highlighting nine SDGs (see fig. 9-2).[3] We believe strongly that each owner's discrete investment decision, in aggregate form, will have significant global consequences.

While volumes are being written about each of the SDGs, we examine, in detail, why we believe in the power of existing buildings to change the world we live in.[4]

No Poverty: The high costs of energy contribute to poverty issues globally. It is suggested that more than 150 million people in the European Union alone are living in "fuel poverty," meaning that more than 10 percent of people's annual income is spent on energy.[5] We see this same fuel poverty happening in cities throughout the United States. There is now a better approach to energy-efficiency retrofits, saving money for occupants and building owners. Retrofitting existing buildings reduces impact on the environment and provides direct and meaning-

Figure 9-2. Where and How Buildings Impact the UN SDGs (used with permission from the World Green Business Council).

ful impact on economic development for this and other SDGs.[6] The Healthy Schools Network's "Ignored Too Long" report, created by a US-based panel of experts from the US Government Accountability Office, found that the poorest children in the poorest communities tended to have school facilities in the worst condition, thus causing children to suffer from the worst environmental exposures.[7]

Good Health and Well-Being: By controlling indoor environmental quality using the approaches outlined in this book, building owners and developers can now affordably set ultra-high indoor air-quality goals and continuously monitor results. According to the US EPA, each year millions of children experience elevated blood levels of contaminants resulting from their exposure to indoor pollutants.[8] Carnegie Mellon University's Center for Building Performance in Pittsburgh, Pennsylvania, reviewed five separate studies evaluating the impact of improved air quality on asthma and found an average reduction in asthma of 38.5 percent in buildings with improved air quality.[9] Benefits of improving indoor environmental quality for children include reductions in asthma-related school absenteeism, improved concentration, and improved performance as measured by grades.[10] Owners of all buildings, especially hospitals and schools, should not focus on replacing failed systems in kind when we know that there are holistic renovation options that save money and contribute to patient recovery and the health and productivity of students, enabling all people to thrive.

Quality Education: Ensuring that every child in every school all over the world has access to quality education has been a topic of great interest and a focus of commitment for nations, states, cities, foundations, and parents. What needs more exposure is the growing body of work indicating the impact a school's indoor environmental quality has on staff and student health and performance. The evidence is overwhelming:

- Studies have shown a 50–70 percent increase of respiratory illness in spaces with low ventilation rates.[11]
- Schools in California reported a 65 percent reduction in asthma cases among elementary students when school indoor environmental quality is improved.[12]
- Students have faster and more accurate responses to cognitive function tests at high ventilation rates.[13]
- Students exposed to daylight attended school 3.2 to 3.8 more days per year.[14]
- Students in daylit environments showed a 20–26 percent improvement on test scores compared to students in traditionally lit environments.[15]

Affordable and Clean Energy: By pursuing the Natural Order of Sustainability, reaching affordable and clean energy is possible if you know when and how to invest. Working to find an existing building's theoretical optimum level of performance while looking at the building holistically provides a path for decision makers to employ renewable-energy solutions in the most affordable manner possible. The low-hanging fruit in existing buildings is typically found in energy conservation. Keep in mind, the cheapest form of energy is the energy we don't use, and conservation is a building block to other integrated building initiatives.

Decent Work and Economic Growth: Existing buildings have the potential to provide healthier indoor environmental quality that leads in turn to improved work environments and economic growth. The reliance on local value chains for buildings and the engagement of small and medium enterprises (SMEs) throughout the life of a building, along with locally made and developed building materials, create jobs and economic growth. A green economy is a growth engine for the United States and elsewhere.[16] A study from Copenhagen Economics in 2012 showed that an annual investment of USD $56B in the energy-efficient renovation of existing buildings in the European Union through 2020 would create approximately 760,000 jobs each year. This would lead to an enduring net annual improvement in public budgets of between USD $41B and USD $56B.[17] Since the existing buildings with energy-management systems offer benefits beyond reduced energy consumption, this smart-building segment of the building industry will grow significantly in the future as we realize the full potential of, and business case for, high-performance buildings, smart communities, and smart cities.[18]

Industry Innovation and Infrastructure: Every time energy-efficiency or sustainability industries take a step forward and prove what's possible, entire cottage industries are born. Climate change now calls for future-proofing buildings so that buildings are resilient and can withstand future risks and shocks to infrastructure. Consider smart buildings: roughly two years ago, the smart-building market was valued at about $5.8M. Recently, *Energy Manager Today* reported that this market will reach almost $62M. Zion Market Research expects this market to exhibit a compounded annual growth rate of more than 34 percent over a seven-year period.[19]

The major factor driving the growth of the smart-building market is the globally increasing concern about energy consumption.[20] The Architecture 2030 Challenge for existing buildings to reduce water and energy consumption 50 percent by 2030 promotes innovation and resilient buildings. Add to this the ability of buildings to find ways to Zero-Energy and zero emissions, and we can see that buildings are significant drivers of new technology and innovation.

Sustainable Cities and Communities: Access to safe, affordable, and healthy housing and reasonable air quality is paramount to creating sustainable cities and communities. It is irrational to think that this goal will be accomplished through new housing around the world. With over 60 percent of the world's population living in urban areas in the near future, it is essential that building owners and developers of multifamily buildings around the world embrace the tools, techniques, and mindset in this book and invest to make old buildings perform like new ones. Further elaborations of this goal will require city policies, homes, offices, schools, and the built environment, in general, to support sustainable building practices across urban areas. Current contributions to this goal can be seen in the development of smart cities and registered living communities within the International Living Future Institute's living community challenge.[21] The tools already exist to ensure that existing buildings are part of symbiotic relationships between people and all aspects of the built environment. Governments, campuses, planners, developers, and neighborhoods are coming together to create connected, symbiotic communities.

Responsible Consumption and Production: In order to minimize the use of energy, to optimize space, and to reduce the environmental impact of buildings, Internet-of-things-enabled smart buildings are the responsible option. Considering all the aspects of smart-building infrastructure provides facility managers and owners improved asset performance. This goal places our attention on resource and energy efficiency, while also supporting other goals to help with sustainable infrastructure, access to services, and jobs. Here, the circular economy and cradle-to-cradle models include existing buildings and their occupants to help ensure minimal or zero waste going to landfills. When involved in renovations of existing buildings, the procurement of recycled, refurbished, and upcycled products within building spaces can contribute to improved indoor environmental air quality. These efforts also support local and regional manufacturing and reduction of GHG emissions from the transportation of waste and reclaimed materials to these production facilities, while avoiding the need for the transportation of products and materials across oceans. Supporting this goal also aligns with United States Green Building Council LEED points for sourcing buildings materials within 500 miles of a building site.

Climate Action: To positively affect climate change, existing buildings must reduce their greenhouse-gas emissions. By reducing dependency on energy, we reduce greenhouse-gas emissions, and the relationship is direct and meaningful. This book provides every owner, developer, and politician the data and strategies they need to aggressively pursue a reduction of greenhouse gas emissions in the built environment. There is no reason that existing buildings all over the world cannot affordably reduce their energy consumption by 50–85 percent. In many major cities like New York City and Seattle, 2.5–3 percent of buildings produce almost half of the area's greenhouse gases. Think about that: we only have to convince 3 percent of building owners to reduce their energy use by about 50 percent in order to reap extraordinary benefits.

Life on Land: The application of common sense and basic science says that we should not extract materials from the earth faster than they can be naturally replenished. The reuse of and investment in existing buildings helps to minimize stress on natural systems for building materials by utilizing the embodied energy inherent in existing buildings. Further, the transportation of materials across supply chains is a major contributor to GHG emissions and climate change. Sustainable supply-chain management and the use of environmental and social sustainability standards, along with product certifications, helps to reduce the negative impacts on land, air, water, and ecosystems.[22] Reducing impacts to life on land can be integrated into any building project during and after construction. Many projects are incorporating ways to enhance biodiversity not only within buildings (i.e., indoor living walls, with research showing that a dynamic microbiological environment is good for human health) but also as part of building structures (i.e., green walls and roofs) and their exteriors.

Partnerships for the Goals: A purposeful part of this book is to push for more collaboration and to advocate for partnerships. Holistic building solutions are the result of the integration and collaboration of architects, engineers, construction professionals. The UN SDGs exist to strengthen global partnerships to support and achieve the ambitious targets for all goals by 2030. This means bringing together national governments, the international community, civil society, the private sector, and other actors to show the business case for existing buildings. Despite advances in certain local, regional, and national areas of the built environment, more needs to be done to accelerate the progress of upgrading existing buildings. In February 2019, at a joint meeting of the United Nations Environment and Global Alliance for Building and Construction in Ottawa, a question was asked of Rob Bernhardt, CEO Passive House Canada. "What is the most significant development for building regulations and for enabling innovation?" His response: "The growing recognition that clearly defined building performance outcomes requiring the highest levels of efficiency, drive innovation and create economic opportunity. Such outcome-based codes combine efficiency and innovation to enable affordable, comfortable, healthy, and resilient buildings."[23]

These efforts and global goals are a call to action for strengthening the means of implementation and revitalization of partnerships for sustainable development. We all need to refocus and intensify our efforts on buildings and technology where progress has been slow. Targets and indicators are in place to track partnerships and progress toward all SDGs. Targets include finance, technology, capacity-building, trade, and policy, along with institutional coherence, all of which are aligned to support sustainable development.

For the first time in human history, there is a concerted global effort to rethink human systems and our interactions with various ecosystems. Manifestations of this global effort are found in the Intergovernmental Panel on Climate Change (IPCC), the annual UN climate change conference (also called the Conference of the Parties, COP), UN SDGs, and the formation of the Global Alliance for Building and Construction (GlobalABC). Buildings have been at the heart of human and economic development for millennia. Existing buildings now play an increasingly important role in delivering evidence of true sustainable development. Buildings can save the world!

Trends Driving Change for Existing-Building Owners

If you are like most of us, it is important for you to keep up with all that is happening in your individual areas of professional expertise and to look for multidisciplinary opportunities and ideas. It can be noted that the difference between success and failure in any industry will be the ability to recognize and implement significant change. Trends driving overall industry objectives for the next decade include global goals for sustainable development, increased urbanization, increased impacts of consumer technology adoption, increased importance of health and well-being, growing consumer concerns about climate change and sustainability, and impacts of next-generation information technologies.[24] Add to this the recent predictions from the IPCC on climate change, toxic air in China reducing life expectancy, and Pope Francis's *Laudato si'* with its calls for maintaining the integrity of natural resources, and we should ask ourselves a few important questions: What kind of decision makers will emerge during these times of change? How will leaders recognize existing buildings as catalysts for change? And how will the power of your building be part of the emerging solutions to local and global challenges? The path to affordably making old buildings perform like new ones, while virtually untrodden today, will one day be well worn. The benefits to people and the planet require that we tamp that path down sooner rather than later.

An integrated approach to unleashing the power of existing buildings, as outlined in this book, sends a clear message to architects, engineers, construction professionals, occupants, and owners:

1. Existing buildings are important!
2. Systems at the end of their life cycles and buildings subjected to deferred maintenance are the triggers to tackle holistic restoration of existing buildings. All things being equal, we should never build new. We should always invest in existing buildings to get them performing like new ones.
3. Done properly, existing-building renovations have the potential to reduce energy consumption, reduce carbon emissions, free people from fuel poverty, and ensure that all people have equitable access to healthy and productive indoor environments.

Conclusion

Knowing what you know now, what actions will you take next?

The status quo will not bring about the innovation necessary to uncover the full potential of your building. The power of existing buildings provides many opportunities for the improvement of integrated bottom-line returns on investments in buildings that save money, improve health, and reduce environmental impacts. The business case to renovate existing buildings is most effectively and confidently achieved using the Natural Order of Sustainability—passive first, active second, renewables last.

Increasingly, we are finding that effective communication is a strategic imperative, especially when trying to convince people to invest in restoring an old building rather than simply razing it and starting over. Using an evidence-based approach and whole-building modeling and simulations increases the confidence of investors. Performance is taken to a new level by recognizing the importance of not dwelling on traditional construction-industry language, but instead converting to performance-based metrics communicated using business language. When this happens, project teams and owners are aligned on mission.

Keep in mind: sustainability means more today than simply greenhouse-gas reduction. Sustainability, interpreted literally, means that we do what's necessary to sustain people and the planet on which we all live. The breakthrough in thinking today is that sustainability goals for the built environment don't have to cost more. It's so elegant in the simple fact that we (1) restore old buildings and make them perform like new, (2) ensure that an owner's financial goals are aligned with meeting occupants' indoor environmental needs, and (3) contribute to the planet's needs for reduced greenhouse gas emissions, reduced energy consumption, and reduced water consumption.

Finally, by measuring performance against goals, we elevate the performance of the entire industry. If we choose not to measure results, then it's back to making transactional decisions and listening to the loudest voice in the room. By measuring results, the numbers speak for themselves and drive experienced building performance advocates to the center of the discussion. The advocates for existing buildings know that today's buildings are the result of investments made in the past. Tomorrow's buildings will be the result of investments and choices made today.

As can be seen from this chapter, and this book, existing buildings are complex. Sustainability, whether we like it or not, is now part of the present and the future for owners, construction teams, occupants, and building managers. Owners of building stock all over the world view many of their existing buildings as bottomless money pits. Industry trends and sustainability aspirations are, in fact, critical building performance opportunities. Sustainability is part of the changing landscape and is affecting not only existing buildings but also businesses within those buildings, their employees, and their competition. As buildings, technology, and sustainable solutions emerge, the only questions left are Are you ready? Is your organization ready? Is your building ready?

This book is just the first step toward action. To get the maximum benefits from your buildings, join professional associations that engage in the development of sustainability in your profession. Talk to and work with sustainability professionals within and outside your organization. In Pittsburgh, Pennsylvania, we have an organization called the Green Building Alliance (GBA). GBA offers hundreds of opportunities every year for sustainability professionals to learn, meet, and engage on our region's most important issues. Find similar organizations like this near you to engage with and learn from.

This is the beginning of the journey for all of us, and it requires a paradigm shift toward true integrated bottom-line performance, enabled by empirical data from technology-driven collaboration for the purpose of proving the massive power of existing buildings to drive the achievement of the most important goals of this century.

The greenest building is the one that's already standing.
— Carl Elefante

Project Development Homework

- ✔ Discuss this book and its approach with colleagues.
- ✔ Find ways to include social cost of carbon (i.e., the negative environmental and social impacts avoided with a dollar-per-ton value) and the United Nations' Sustainability Develop-

ment Goals (UN SDGs) in existing building goals and strategic planning.

✔ Map your existing-building projects onto corresponding UN
SDGs to find the new value that your building brings to community and global issues. Then integrate these goals and the
benefits of your project when you pitch it to decision makers.

✔ Recruit others on your team as part of a call to action for existing-building advocates to show the power of existing buildings as catalysts for change and value-creation.

Appendix 1 — Building Your Plan: Project Development Homework

What follows are the "Project Development Homework" lists from each chapter. Together they create a checklist for approaching the important process of renovating existing buildings. Having a guide can increase the likelihood that you, your team, and an existing building reach your performance and operational goals.

Chapter 1: My Building Has High-Performance Potential

✔ Identify a building/project you know of that is in dire need of renewal. Ideally the building serves essential purposes and could never be considered "disposable" to the community in which exists.

✔ Start thinking about how and why you want this existing building to be high-performance.

✔ Think about how transforming the building aligns with the values of the owner. In narrative form, write out how you would articulate this option to the building's owners. You may have to make assumptions around the organization's values, but that's okay.

✔ Think about what the owner has told you is important about the building and that will enable you to begin getting a sense of their basic values.

✔ Storyboard your narrative as a successful progression of your project/building through time to a future state of becoming a high-performance building.

Chapter 2: Where Do I Start?

✔ Consider the building you identified at the end of chapter 1. What would you say are the greatest opportunities for that building if it were able to be transformed into a high-performance building?

✔ What is the current performance of the existing building in terms of empirical data? This baseline information is necessary to begin planning.

✔ How might the owners of the building benefit if that building's performance were to become best-in-its-class?

✔ If you were the owner of that building, what would help you gain the confidence necessary to make the proper investment in the building?

Chapter 3: The Importance of a Project Plan: Every Building Needs One

✔ Using the building/project you have identified so far, what specific goals would you set relative to energy performance, indoor air and environmental quality, first costs, and/or long-term operating costs?

✔ Can you break down whole-building goals or project goals by area of function of the building?

✔ If you can achieve those goals, who benefits from each goal?

Chapter 4: Can I Afford This?

Consider a traditional first-cost approach to any planned renovation to analyze against a whole-building, deep-energy retrofit approach. Then, take a step back and look for all the benefits your renovation can provide, and include this valuation in your ROI, NPV, and TCO. Can you stretch your goals even further to a zero-energy solution, or a lowest-energy approach to the integrated design?

✔ How much money can you save if you transition from a first-cost, cost-savings approach to the project to a cost-avoidance and value-maximization approach?

✔ Identify and rank the pitfalls and barriers you may encounter when developing and implementing your whole-building design.

✔ Run a design-charrette / scenario-planning exercise with your team. What are the worst-case and best-case scenarios for your existing-building project? Acknowledge the risks, and then capture the long-term integrated financial, environmental, and social benefits of the project.

✔ If you know how much your energy consumption can be reduced, find the tonnage equivalent of CO_2e and apply social-cost-of-carbon (SCC) calculations to show the impacts avoided, and thus the value created, by your project, which takes you beyond a single bottom-line, first-costs approach to decision-making.

Chapter 5: The Envelope Holds the Key

✔ Consider the envelope conditions of your target project. Specifically, focus on the fenestration, thermal barrier, air barrier and air leakage, and thermal bridging.

✔ What changes would you instinctively make to improve the overall performance of the envelope in terms of energy consumption and indoor environmental and air quality?

✔ Try to calculate the impact of those improvements on long-term operating costs, human health, and the productivity of your building occupants.

Chapter 6: How Realistic Is Zero-Energy for an Old Building?

How do you define Zero-Energy?

✔ Break your zero-energy opportunity down into subsystems.
  • Which subsystems present the best business case for change?

✔ Where is a logical place to start for your building?

✔ Can you create a master plan that can be implemented in phases in order to take advantage of natural triggers of life cycle, deferred maintenance, renovations, and other conditions to reach zero-energy over a period of time?

✔ If you cannot get to zero, how low can you go?

Chapter 7: Operating Buildings for High Impact

✔ What are the building systems (meters and sensors) that can provide feedback for critical key performance indicators necessary to demonstrate the conversion of your existing building to a high-performance building?

✔ What indoor air-quality sensors will enable you to prove success to stakeholders? Are these sensors reliable and can they be easily recalibrated to ensure continuous data reliability?

✔ Write up a brief description of *how* repurposing performance and building meters and sensors will build the business case to extract integrated value in your existing-building project. Use this information to tell a compelling story about *why* investments in your building are necessary.

✔ Develop a mental model of your future dashboard and sketch it on paper or on a tablet.

✔ Identify the data necessary to operate high-performance buildings.

Chapter 8: Case Studies

✔ Review these cases in light of your own building(s). What strategies and ideas can you reuse in your planned project? Where do you start? How will you go about planning? Can you afford this? What will you do about the envelope? How will you operate the building for high performance? Remember that each step offers a real opportunity to build upon the innovative ideas of others to transform your building.

Chapter 9: Existing Buildings Can Save the World

✔ Discuss this book and its approach with colleagues.

✔ Find ways to include social cost of carbon (i.e., the negative environmental and social impacts avoided with a dollar-per-ton value) and the United Nations' Sustainability Development Goals (UN SDGs) in existing-building goals and strategic planning.

✔ Map your existing building projects onto corresponding UN SDGs to find the new value that your building brings to community and global issues. Then integrate these goals and the benefits of your project when you pitch it to decision makers.

✔ Recruit others on your team as part of a call to action for existing-building advocates to show the power of existing buildings as catalysts for change and value-creation.

Appendix 2 — Critical Resources
on Existing Buildings

The following articles, white papers, journal publications, websites, and books are indispensable resources that will help you quickly get up to speed on a given topic.

Indoor Air Quality

Colton, Meryl D., Piers MacNaughton, Jose Vallarino, John Kane, Mae Bennett-Fripp, John D. Spengler, and Gary Adamkiewicz. "Indoor Air Quality in Green vs. Conventional Multifamily Low-Income Housing." *Environmental Science & Technology* 48, no. 14 (2014): 7833–41.

Darcin, Polat, and Ayse Balanli. "Building User in Terms of Indoor Air-Pollution Exposure," Turkey, Dicle University 1st International Architecture Symposium, 2018. https://www.academia.edu/38194846/Building_User_in_terms_of_Indoor_Air_Pollution_Exposure.

Energy Efficiency

Brew, James Scott. "Achieving Passivhaus Standard in North America: Lessons Learned." *ASHRAE Transactions* 117, no. 2 (2011).

The City of New York, Mayor Bill de Blasio, "New York City's Green New Deal," 2019, https://onenyc.cityofnewyork.us/. This pledge commits New York to carbon neutrality by 2050, and mandates that existing buildings implement retrofits to be more efficient while cutting emissions.

Far, Claire, and Harry Far. "Improving Energy Efficiency of Existing Residential Buildings Using Effective Thermal Retrofit of Building Envelope." *Indoor and Built Environment* (August 22, 2018). doi/10.1177/1420326X18794010.

International Energy Agency. "Capturing the Multiple Benefits of Energy Efficiency." Paris: OECD/IEA, 2014. https://www.gob.mx/cms/uploads /attachment/file/85309/Bibliograf_a_10.pdf.

Next 10. "Untapped Potential of Commercial Buildings: Energy Use and Emissions," (2009). https://next10.org/sites/default/files/NXT10 _BuildingEfficiencies_final.pdf.

Rocky Mountain Institute. "The Case for Net-Zero Energy in the GSA's Owned Building Portfolio" (White Paper.) Washington, DC: General Services Administration, 2015. https://www.gsa.gov/cdnstatic/RMI _white_paper_-_GSA_NZE-_2015-10-21.pdf.

World Resources Institute. "Accelerating Building Efficiency: Eight Actions for Urban Leaders." Washington, DC: WRI, 2019. ISBN 978-1-56973-889-4. https://publications.wri.org/buildingefficiency/

Carbon Emissions

Frey, Patrice, Liz Dunn, Ric Cochran, Katie Spataro, Jason F. McLennan, Ralph DiNola, Nina Tallering, et al. "The Greenest Building: Quantifying the Environmental Value of Building Reuse." Preservation Green Lab, National Trust for Historic Preservation, 2011.

Hawken, Paul, ed. *Drawdown: The Most Comprehensive Plan Ever Proposed to Reverse Global Warming: 100 Solutions to Reverse Global Warming.* London: Penguin, 2017.

Preservation Green Lab. "The Greenest Building: Quantifying the Environmental Value of Building Reuse." National Trust for Historic Preservation. 2011. https://living-future.org/wp-content/uploads/2016/11 /The_Greenest_Building.pdf.

United States Environmental Protection Agency. "The Social Cost of Carbon: Estimating the Benefits of Reducing Greenhouse-Gas Emissions." *Climate Change* (website). EPA, 2017. https://19january2017snapshot .epa.gov/climatechange/social-cost-carbon_.html.

Sustainable Development

Hawken, Paul, Amory B. Lovins, and L. Hunter Lovins. *Natural Capitalism: The Next Industrial Revolution.* London: Routledge, 2013.

Lovins, Amory. *Reinventing Fire: Bold Business Solutions for the New Energy Era.* White River Junction, VT: Chelsea Green Publishing, 2013.

Sroufe, Robert P. *Integrated Management: How Sustainability Creates Value for Any Business*. Bingley, UK: Emerald Publishing Limited, 2018.

United Nations. "Sustainable Development Goals." *Knowledge Platform* (website). https://sustainabledevelopment.un.org/?menu=1300.

Wines, James, and Philip Jodidio. *Green Architecture*. Köln: Taschen, 2000.

Business Case for Existing Buildings

International Living Future Institute. "High-Performance Green Building: What's It Worth?" May 2009. https://living-future.org/wp-content /uploads/2016/11/High_Performance_Green_Building.pdf.

Jungclaus, Matt, Alisa Petersen, and Cara Carmichael. *Guide: Best Practices for Achieving Zero Over Time For Buildings Portfolio*. Rocky Mountain Institute and the Urban Land Institute, 2018. https://www.rmi.org /wp-content/uploads/2018/09/RMI_Best_Practices_for_Achieving _Zero_Over_Time_2018.pdf.

McGraw Hill Construction. "Business Case for Energy-Efficient Building: Retrofit and Renovation." https://www.energy.gov/sites/prod/files/2013 /12/f5/business_case_for_energy_efficiency_retrofit_renovation _smr_2011.pdf.

New Buildings Institute. "A Search for Deep Energy Savings in Existing Buildings." Preservation Green Lab, National Trust for Historic Preservation. September 2011. http://www.josre.org/wp-content/uploads /2012/10/11DeepSavingsEBCaseStudiesNBI.pdf.

Rocky Mountain Institute. "Buildings." 2019. https://rmi.org/our-work /buildings/.

———. "The Case for Net-Zero Energy in the GSA's Owned Building Portfolio." White Paper. October 16, 2015. https://www.gsa.gov /cdnstatic/RMI_white_paper_-_GSA_NZE-_2015-10-21.pdf.

Stok. "The Financial Case for High Performance Buildings: Quantifying the Bottom Line of Improved Productivity, Retention, and Wellness." 2018. https://stok.com/research/financial-case-for-high-performance -buildings.

US Department of Energy. "The Business Case for High Performance-Buildings." *Better Buildings* (website), 2017. https://betterbuildings solutiocenter.energy.gov/videos/business-case-high-performance -buildings.

———. "Making the Business Case for High-Performance Buildings." Office of Energy Efficiency and Renewable Energy, November 23, 2017. https://www.energy.gov/eere/buildings/articles/making-business -case-high-performance-buildings.

———. "Making the Business Case for High Performance Green Buildings." Office of Energy Efficiency and Renewable Energy, November 21, 2016. https://www.energystar.gov/buildings/tools-and-resources /making-business-case-high-performance-green-buildings.

US Environmental Protection Agency. "Making the Case for High-Performance Green Buildings." *Energy Star* (website), 2016. https://www.energystar.gov/buildings/tools-and-resources/making -business-case-high-performance-green-buildings.

US Green Building Council. "The Business Case for Green Building." February 10, 2015. https://www.usgbc.org/articles/business-case-green -building.

Vancouver Economic Commission. "Green Buildings Market Forecast: Demand for Building Products, Metro Vancouver, 2019–2032." March 2019. http://www.vancouvereconomic.com/wp-content/uploads/2019 /03/GreenBuildingsMarketResearch_WEBMarch7_Launch-compressed -compressed.pdf.

World Green Building Council. "Building the Business Case: Health, Wellbeing and Productivity in Green Offices." *Better Places for People* (website), October 2016. https://www.worldgbc.org/sites/default /files/WGBC_BtBC_Dec2016_Digital_Low-MAY24_0.pdf.

———. "New WorldGBC Report Makes Business Case for Building Green with Health and Wellbeing Features." April 24, 2018. https://www .worldgbc.org/news-media/new-worldgbc-report-makes-business -case-building-green-with-health-and-wellbeing-features.

Schools

Goodson, Dana, and Kim Rustem. "Environmental Health at School: Ignored Too Long." ERIC, Healthy Schools Network, Panel and Facilitated Workshop, Summary Report, 2015. https://eric.ed.gov /?id=ED570456.

Healthy Schools Network, Inc. "Daylighting." 2012. http://www.healthy schools.org/data/files/Daylighting.pdf.

Heschong Mahone Group. "Daylighting in Schools: An Investigation into the Relationship Between Daylighting and Human Performance." Condensed Report, August 20, 1999. https://www.pge .com/includes/docs/pdfs/shared/edusafety/training/pec/daylight /SchoolsCondensed820.pdf.

Kats, Gregory. "Greening America's Schools—Costs and Benefits." Capital E Report, October 2006. https://www.usgbc.org/sites/default/files /Greening_Americas_Schools.pdf.

US General Accounting Office. *School Facilities: Condition of America's Schools.* Report to Congressional Requesters, February 1995. https://www.gao.gov/assets/230/220864.pdf.

Cognitive Function

Harvard University. "The Impact of Green Buildings on Cognitive Function." *Sustainability* (website), 2018. https://green.harvard.edu /tools-resources/research-highlight/impact-green-buildings-cognitive -function.

MacNaughton, Piers, Usha Satish, Jose Guillermo Cedeno Laurent, Skye Flanigan, Jose Vallarino, Brent Coull, John D. Spengler, and Joseph G. Allen. "The Impact of Working in a Green-Certified Building on Cognitive Function and Health." *Building and Environment* 114 (2017): 178–86.

Organizations

American Institute of Architects: https://www.aia.org/

Architecture 2030: https://architecture2030.org/

Better Buildings Initiative, US Department of Energy: https://betterbuild ingssolutioncenter.energy.gov/

International Living Future Institute: https://living-future.org/

International WELL Building Institute: https://www.wellcertified.com/en

Passive House Institute US (PHIUS): http://www.phius.org/home-page

RESET Healthy Buildings Standard: https://www.reset.build/

Rocky Mountain Institute: https://www.rmi.org/

United Nations Sustainable Development Goals: https://sustainable development.un.org/?menu=1300

US Green Building Council: https://new.usgbc.org/

Notes

Introduction

1. C. Elefante, "The Greenest Building Is . . . One That Is Already Built," *Forum Journal, The Journal of the National Trust for Historic Preservation* 21, no. 4 (2007): 26–28.
2. See: Intergovernmental Panel on Climate Change, https://www.ipcc .ch/; and see: Architecture 2030 and the 2030 Challenge, along with other work by this organization, https://architecture2030.org/.
3. Architecture 2030, "Buildings Generate Nearly 40% of Annual Global GHG Emissions," accessed February 23, 2019, https://architecture 2030.org/buildings_problem_why/.
4. B. Heather, World Green Building Trends in 2016: Business Benefits, accessed February 23, 2019, http://www.usgbc.org/articles/world -green-building-trends-2016-business-benefit
5. Mike Jackson, "Embodied Energy and Historic Preservation: A Needed Reassessment," *APT Bulletin* 36, no. 4, (2005): 47–52.

Chapter 1

1. G. MacKenzie, *Orbiting the Giant Hairball: A Corporate Fool's Guide to Surviving with Grace* (New York: Viking Press, 1998).
2. D. Meadows, *Thinking in Systems: A Primer* (White River Junction, VT: Greenleaf Publishing, 2008), 327.
3. City of New York, Mayor Bill de Blasio's Office of Sustainability, "Inventory of New York City Greenhouse Gas Emissions in 2015," https://www.dec.ny.gov/docs/administration_pdf/nycghg.pdf.
4. Mayor's Press Office, "Action on Global Warming: NYC's Green New Deal," April 22, 2019, https://www1.nyc.gov/office-of-the-mayor /news/209-19/action-global-warming-nyc-s-green-new-deal#/0.
5. K. Haanaes et al., "Sustainability: The 'Embracers' Seize Advantage"

(Boston: MIT Sloan and Boston Consulting Group, 2011), https://sloanreview.mit.edu/projects/sustainability-the-embracers-seize-advantage/.

6. E. Tidhar et al., "Toward the Next Horizon of Industry 4.0—Building Collaboration through Collaborations and Startups" (Deloitte Insights, 2018), accessed February 23, 2019, https://www2.deloitte.com/insights/us/en/focus/industry-4-0/building-capabilities-through-collaborations-startups.html.

Chapter 2

1. Z. Bakó-Biró et al., "Ventilation Rates in Schools and Learning Performance," *Proceedings of the CLIMA 2007 WellBeing Indoors* (2007). 1434–40.

2. N. Klepeis et al. "The National Human Activity Pattern Survey (NHAPS): A Resource for Assessing Exposure to Environmental Pollutants, *Journal of Exposure Analysis and Environmental Epidemiology* 11, no. 3 (2001): 231–52, https://www.nature.com/articles/7500165.

3. Environmental Protection Agency, EPA Report to Congress on Indoor Air Quality, vol. I—Federal Programs Addressing Indoor Air Quality, EPA/400/1-89.001B (Washington, DC: US Government Printing Office, 1989).

Chapter 3

1. *Charrette* is a French word for cart or chariot. A charrette was used in the nineteenth century in Paris to shuttle architecture students between their works prior to a deadline. The students would continue working *en charrette*, or "in the cart," right up until the deadline. See: Patrick Henry Winston. "En Charrette," *Slice of MIT* (blog), May 2010.

2. G. Lindsey, J. A. Todd, and S. J. Hayter, *A Handbook for Planning and Conducting Charettes for High-Performance Projects*, NREL/BK-710-33425, National Renewable Energy Laboratory (August 2003), https://www.nrel.gov/docs/fy03osti/33425.pdf.

3. See: Zimmer Gunsul Frasca Architects LLP, ZGF publication from Rocky Mountain Institute's Innovation Center, www.zgf.com/project/rmi.

4. These insights are based on our own observations of the Innovation Center along with secondary sources of information.

Chapter 4

1. Quoted by R. Clarke et al., "The Challenge of Going Green," *Harvard Business Review* (July-August 1994), accessed February 23, 2019, https://hbr.org/1994/07/the-challenge-of-going-green.
2. Ed Mazria, presentation at Penn State Energy Days (May 2018).
3. Deloitte, "Breakthrough for Sustainability in Commercial Real Estate", The Deloitte Center for Financial Services, 2014.
4. See for example, World Green Building Council's Report, "Building the Business Case: Health, Wellbeing and Productivity in Green Offices", October, 2016.
5. See: Integrated Project Delivery, accessed at http://info.aia.org/site objects/files/ipd_guide_2007.pdf.
6. For information on "tunneling through costs," see: A. Lovins, H. Lovins, and P. Hawken, *Natural Capitalism* (Boston: Little, Brown and Company, 2000), chap. 6.
7. See: Environmental Protection Agency, "The Social Cost of Carbon: Estimating the Benefits of Reducing Greenhouse-Gas Emissions" (2017), https://19january2017snapshot.epa.gov/climatechange/social-cost -carbon_.html; see also: Environmental Defense Fund, "The True Cost of Carbon Pollution: How the Social Cost of Carbon Improves Policies to Address Climate Change" (2019), https://www.edf.org /true-cost-carbon-pollution. Corporations have been using this for years, with the most common value being $40 per ton; projections are for this to be $100 per ton after 2020.
8. Heather Clancy, "Wal-Mart, Disney, Microsoft Hedge Bets on Carbon Pricing, *GreenBiz* (December 6, 2013), accessed March 20, 2019, https://www.greenbiz.com/blog/2013/12/06/us-companies-hedge -bets-against-future-regulation-internal-carbon-pricing.

Chapter 5

1. Laura Legere, "Built-In Green Energy—How a PA Affordable Housing Agency Is Making Ultra-Efficient Buildings Mainstream," *Pittsburgh Post-Gazette*, December 30, 2018.
2. See, for example: G. Hovnanian and Erik Sjödin, "How Analytics Can Drive Smarter Engineering and Construction Decisions," McKinsey & Company, Capital Projects & Infrastructure (January 2019).
3. Harvard Center for Health and the Global Environment, "The Impact of Green Buildings on Cognitive Function," *Harvard University—*

Sustainability (2018), accessed February 32, 2019, https://green .harvard.edu/tools-resources/research-highlight/impact-green -buildings-cognitive-function.

4. Harvard School of Public Health, "Nine Foundations of a Healthy Building Project," *For Health* (blog), accessed February 23, 2019, https://9foundations.forhealth.org/; see also: J. Cvox-Ganser, "Exposure to Residential Dampness and Mold Contributed to 21% of 21.8 Million Cases of Asthma Each Year," *PubMed Central*, National Institutes of Health (2019), accessed February 23, 2019, https://www.ncbi .nlm.nih.gov/pmc/articles/PMC4667360/.

5. "Foundations for Student Success: How School Buildings Influence Student Health, Thinking, and Performance," Harvard School of Public Health, *For Health*, accessed February 23, 2019 https://forhealth.org/Harvard.Schools_For_Health.Foundations_ for_Student_Success.pdf; see also: Centers for Disease Control and Prevention, "Asthma-Related Missed School Days among Children Aged 5–17 Years," US Department of Health and Human Services (October 5, 2015), https://www.cdc.gov/asthma/asthma_stats /missing_days.htm.

Chapter 6

1. Preservation Green Lab, "The Greenest Building: Quantifying the Environmental Value of Building Reuse" (National Trust for Historical Preservation, 2011), https://living-future.org/wp-content /uploads/2016/11/The_Greenest_Building.pdf.

2. Building Efficiency Initiative, "Why Focus on Existing Buildings?" News Article, April, 2010.

3. Mayor's Office of Long-Term Planning and Sustainability, "One City: Built to Last," City of New York, accessed March 28, 2019, http://www .nyc.gov/html/builttolast/assets/downloads/pdf/OneCity.pdf.

4. S. Pless and P. Torcellini, "Net-Zero Energy Building: A Classification System Based on Renewable Energy-Supply Options," National Renewable Energy Laboratory (2010), accessed March 28, 2019, https://www.nrel.gov/docs/fy10osti/44586.pdf.

5. P. Torcellini, S. Pless, M. Deru, and D. Crawley, "Zero Energy Buildings: A Critical Look at the Definition," National Renewable Energy Laboratory (Reprint, 2006), accessed March 28, 2019, https://www.nrel .gov/docs/fy06osti/39833.pdf.

6. See, for example, Executive Order 13514 from President Barack Obama, requiring federal agencies to measure and reduce greenhouse-gas pollution resulting from federal operations, and to improve energy efficiency, increase the use of renewable energy, reduce water consumption, and purchase energy-efficient and environmentally preferable goods and materials. EO 13514 Archive, accessed March 20, 2019, https://www.fedcenter.gov/programs/eo13514/.

7. General Services Administration (GSA), "Sustainable Facilities Tool—Net Zero Examples," accessed March 20, 2019, https://sftool.gov/learn/about/422/net-examples.

8. The Zero Energy Project, "Zero Energy—Building Case Studies," accessed March 20, 2019, https://zeroenergyproject.org/zero-energy-building-case-studies/.

9. A. Mardiana and S. B. Riffat, "Building Energy Consumption and Carbon Dioxide Emissions: Threat to Climate Change," *Journal of Earth Science and Climatic Change* S3:001 (2015), doi:10.4172/2157-7617.S3-001.

Chapter 7

1. See, for example: "Niagara Framework Guide Specifications to Include Smart Buildings, Tridium," accessed March 28, 2019, www.vykon.com/library/Niagara_4_Guide_Spec.docx.

2. Billy Grayson, "Sustainability in the Age of Big Data," Urbanland (October 29, 2018), https://urbanland.uli.org/sustainability/sustainability-in-the-age-of-big-data/.

3. See: "Gartner Says 8.4 Billion Connected 'Things' Will Be in Use in 2017, Up 31 Percent from 2016," Gartner, Inc., press release, February 7, 2017, accessed February 23, 2019, https://www.gartner.com/en/newsroom/press-releases/2017-02-07-gartner-says-8-billion-connected-things-will-be-in-use-in-2017-up-31-percent-from-2016; see also: L. Columbus, "2017 Roundup of the Internet of Things," *Forbes* (December 10, 2017), accessed January 19, 2019, https://www.forbes.com/sites/louiscolumbus/2017/12/10/2017-roundup-of-internet-of-things-forecasts/#742325461480.

4. "75 Billion IoT Devices Predicted by 2025," *The News*, April 22, 2016, accessed February 15, 2019, https://www.achrnews.com/articles/132303-billion-iot-devices-predicted-by--.

5. A Niagara platform is a hub or central location for Smart Building Infrastructure, including meters and sensors. Its open infrastructure should transparently and securely communicate with disparate, independent, and proprietary systems in the building.

   The Niagara framework facilitates an open, no-lock-in framework architecture, allowing multi-vendor systems and solutions to be connected and supported by a community of systems integrators and developers, allowing clients the freedom of choice either to retain existing systems and infrastructures or to upgrade in the future using the latest technologies and infrastructures. The Niagara framework was designed to allow integrators and developers to connect, manage, and control any device, regardless of manufacturer, using any protocol. Source: Niagara Framework Guide Specifications to Include Smart Buildings. See: "Niagara Framework Guide Specification to Include Smart Buildings," Tridium Inc., 2019, https://www.tridium.com/~/media/tridium/library/documents/niagara%20framework%20middleware%20specification.ashx?la=en.

6. International Energy Agency, *Capturing the Multiple Benefits of Energy Efficiency* (Paris: IEA, 2014).

7. R. Sroufe, *Integrated Management: How Sustainability Creates Value for Any Business* (Bingley, UK: Emerald Press, 2018).

8. L. Gilbert, "WeLL @ Work: The Benefits of a WELL Certified Office," International WELL Building Institute (2017).

9. J. Folger, "The Causes and Costs of Absenteeism," *Investopedia* (April 16, 2018).

10. See: "Reducing Absenteeism in the Workplace, *Employee Benefits* (April 16, 2018), accessed February 19, 2019, https://www.employeebenefits.co.uk/reducing-absenteeism-workplace/; for more information on how good building design increases productivity and reduces medical costs, see: D. J. Clements-Croome, "Sustainable Intelligent Buildings for Better Health, Comfort, and Well-Being: A Report," Denzero International Conference (October 2014), https://www.researchgate.net/publication/266739139_Sustainable_Intelligent_Buildings_for_Better_Health_Comfort_and_Well-being----a_Report.

11. G. Kats, "Greening of America's Schools: Costs and Benefits," *Capital E* (2016), accessed April 15, 2019, https://www.usgbc.org/resources/greening-america039s-schools-costs-and-benefits.

12. R. Srivastava, "Integrating Financial, Environmental, and Human Capital—the Triple Bottom Line—for High-Performance Investments in the Built Environment," PhD Thesis, Carnegie Mellon University, Pittsburgh, PA (2018).

Chapter 8

1. For more on integrated measures of financial performance, see: R. Sroufe, *Integrated Management: How Sustainability Creates Value for Any Business* (Bingley, UK: Emerald Press, 2018).

Chapter 9

1. P. Hawken, A. Lovins, and H. Lovins, *Natural Capitalism*, 10th Anniversary Edition (London: Earthscan from Routledge, 2010).
2. For information on how Preservation Green Lab found this possible for Multnomah County in Washington State, see their 2011 report: "The Greenest Building: Quantifying the Environmental Value of Building Reuse," p. 85.
3. See: "Green Building: Improving the Lives of Billions by Helping to Achieve the UN Sustainable Development Goals," World Green Buildings Council (2017), accessed February 15, 2019, https://www.worldgbc.org/news-media/green-building-improving-lives-billions-helping-achieve-un-sustainable-development-goals.
4. For more on the challenges of improving lives and achieving the Sustainable Development Goals through design and construction in order to help orient your work and your existing building, see: B. Dimson, "Principles and Challenges of Sustainable Design and Construction, UNEP," *Industry and Environment* 19, no. 2 (April-June 1996).
5. International Energy Agency, "Capturing the Multiple Benefits of Energy Efficiency" (2014), accessed February 19, 2019, https://www.iea.org/publications/freepublications/publication/Multiple_Benefits_of_Energy_Efficiency.pdf.
6. See, for example: E. B. Barbier, *A Global Green New Deal: Rethinking the Economic Recovery* (Cambridge Press, 2010); see also: International Energy Agency, *Capturing the Multiple Benefits of Energy Efficiency* (Paris: IEA, 2014).
7. Healthy Schools Network, "Environmental Health at School: Ignored Too Long" (2015), accessed February 23, 2019, https://files.eric.ed.gov/fulltext/ED570456.pdf.

8. US Environmental Protection Agency, "Healthy Buildings, Healthy People: A Vision for the 21st Century," EPA 402-K-01-003 (Washington, DC: US EPA, 2001).

9. G. A. Kats, "Greening America's Schools: Costs and Benefits," *Capital E* (2016), p. 13, accessed April 15, 2019, https://www.usgbc.org /resources/greening-america039s-schools-costs-and-benefits.

10. Y. Y. Meng, S. H. Babey, and J. Wolstein, "Asthma-Related School Absenteeism and School Concentration of Low-Income Students in California," *Preventing Chronic Disease* 9 (2012): E98.

11. R. Loveland, "Dreaming the Future: How Zero-Energy Design Can Transform the School Environment," *Green Schools Catalyst Quarterly* 3 (2017), http://catalyst.greenschoolsnationalnetwork.org/gscatalyst /september_2017/MobilePagedArticle.action?articleId=1152429 #articleId1152429.

12. Meng et al., ibid.

13. Z. Bako-Biro, N. Kochhar, D. J. Clements-Croome, H. B. Awbi, and M. Williams, "Ventilation Rates in Schools and Learning Performance," *Proceedings of CLIMA 2007—WellBeing Indoors* (2007), accessed February 23, 2019, https://www.researchgate.net/publication/242261403 _Ventilation_Rates_in_Schools_and_Learning_Performance.

14. "Daylighting," Healthy Schools Network, Inc. (2012). http://www .healthyschools.org/data/files/Daylighting.pdf.

15. Hershong Mahone Group, "Daylighting in Schools: An Investigation into the Relationship Between Daylighting and Human Performance," Pacific Gas and Electric Company (1999), accessed February 23, 2019, http://h-m-g.com/downloads/Daylighint /schoolc.pdf.

16. United States Green Building Council, "Market Brief: A Green Economy Is a Growth Economy" (2011).

17. See: IEA, "Capturing the Multiple Benefits of Energy Efficiency"; see also: Barbier, *A Global Green New Deal*.

18. Better Buildings Solutions Center, "The Business Case for High-Performance Buildings," US Department of Energy (2017), accessed February 16, 2019, https://betterbuildingssolutioncenter.energy.gov /videos/business-case-high-performance-buildings.

19. "Global Forecast for Smart Building Market Size & Share to Surpass USD 61,900 Million by 2024: Zion Market Research," *Market Watch* (2018), accessed February 16, 2019, https://www.market watch.com/press-release/global-forecast-for-smart-building-market

-size-share-to-surpass-usd-61900-million-by-2024-zion-market
-research-2018-05-25.

20. E. Hollbrook, "Report: Smart Buildings Market to Reach $62 Million by 2024," *Energy Manager Today* (January 10, 2019).

21. For information on the ILFI's living community challenge, see: https: //living-future.org/lcc/.

22. R. Sroufe and S. Melnyk, *Developing Sustainable Supply Chains to Drive Value*, vol. II, *Management Issues, Insight, Concepts, and Tools—Implementation* (New York: Business Expert Press, 2017).

23. Personal communication with Rob Bernhardt, United Nations Environment and Global Alliance for Building and Construction in Ottawa, February 21, 2019.

24. Adopted from Sroufe and Melnyk, *Developing Sustainable Supply Chains*.